HEARING

HEARING

BY

ROBERT MORRIS OGDEN
PROFESSOR OF EDUCATION IN CORNELL UNIVERSITY

WITH DIAGRAMS

NEW YORK
HARCOURT, BRACE AND COMPANY

PRINTED IN THE U. S. A. BY
THE QUINN & BODEN COMPANY
RAHWAY, N. J.

1581616

To

EDWARD BRADFORD TITCHENER

The expense of publishing this volume was in part borne by a grant from the Heckscher Foundation for the Advancement of Research, established by August Heckscher at Cornell University.

PREFACE

This book is a by-product of reviews and critical studies
of the psychology of hearing, a subject with which I have
been engaged over a period dating from my contact with
Professor Max F. Meyer—a distinguished investigator of
audition—while serving as his assistant in the University of
Missouri, 1903-1905. There I began my reviews, for publi-
cation chiefly in *The Psychological Bulletin*. Later, as a
member of the co-operating board of editors of the *Bulletin*,
I took charge of the department dealing with the sense of
hearing, and during the ten succeeding years published an
annual summary of work done in this field (*80*),[1] as well as
numerous more extended reviews and several articles.

My growing interest in the subject finally led me to the
task of bringing together into systematic order the chief
experimental results which have so rapidly accumulated in
recent years. This task I have tried to accomplish in the
present volume, and at the same time to supply from earlier
writings the more generally accepted facts about hearing.
But to bring these widely scattered data within the scope
of a small book was not always easy; for the experimental
results were sometimes conflicting, and the interpretations
that have been placed upon them show a wide variance

[1] The number in parenthesis refers to the exact bibliographical description in the list at the end of the book. Other bibliographical references are similarly provided throughout the body of the book.

of opinion. Therefore, I have worked slowly since under-
taking the first draft of the book in 1916. In 1922, with the
aid of a special grant from the Foundation for the Advancement of Research generously established at Cornell University by Mr. August Heckscher, I was relieved of teaching for a term in order that I might find time thoroughly to
revise my manuscript and to put my book into shape for
the printer.

What I have done is to select those facts which appeared
the more significant and trustworthy, and then allow them
to fall into their own system of organization. Although
I have probably not succeeded in treating these facts
with the complete impartiality that my procedure suggests,
at least I have tried to preserve an open mind on all de-
batable issues, and thus to leave the way clear for new
research to alter and amend my conclusions.

In addition to the organization of facts, I have had constantly in mind the varied interests of my readers. As first
drafted the book was intended as a brief summary, more
or less popular in nature; but, as the work went on, I was
led to amplify my material, and, I fear, to employ a somewhat condensed treatment of many details. Yet it is my
hope that all who take an interest in the sense of hearing
may still find the book clear and informing. Although I
have been unable to avoid the terminology and the attitude which a student of psychology is apt to adopt, I should
be gratified if the teacher, the musician, the linguist, and
the medical practitioner—especially if he be an otologist

—could each find something to help him in the solution of his special problems.

My indebtedness to various persons is so great that I am at a loss to know where to begin or end my acknowledgment. In the course of the years which have passed since I began to collect and to organize the results of research in audition, my correspondence with different investigators has been considerable, and from each I have gained insight and factual data which it would be very difficult to indicate in detail. I can therefore only make this general acknowledgment of my obligations, with the remark that my list of references includes the names of many who have rendered me much personal assistance apart from their published writings.

There are, however, several individuals whose aid has been more intimate. Special sections of the book have been read by my colleagues, Professors H. P. Weld and F. K. Richtmyer, with many helpful comments. The medical aspects of the chapter on the Pathology of Hearing have been checked by Dr. Robert Sonnenschein of Chicago, who kindly offered his services in a field where I must confess myself distinctly an amateur. Finally, the whole manuscript has been greatly improved by the critical insight of my friends and colleagues, Professors Lane Cooper, James T. Quarles, and Edward Bradford Titchener. To the last of these I owe the greatest debt of all; for it was he who first introduced me to the science of psychology, and it is he who still serves me as critic and mentor: wherefore I take the liberty of inscribing my book to him.

R. M. O.

CONTENTS

xi

HEARING

CHAPTER I

PHYSICAL SOUND [1]

How do we hear? And what do we hear? These questions can be answered only by inquiries into three different fields of science—physics, physiology, and psychology. The sound we hear is physical, but it needs the physiological function of the ear to receive it and communicate it to the brain. A third aspect of hearing is the consciousness of sound, the phenomena of which lie in the province of psychology. In this book it is with psychological phenomena that we shall be chiefly concerned. However, it is important to notice that the stimulation of sound is an objective datum of physics—that sound is generated by a vibrating body, and transmitted through an appropriate medium to the sense-organ of hearing. It is important also to have clearly in mind that the functions of the ear, and the nervous excitations accompanying these functions, are distinct problems of a living organism. And hence, before we can proceed to a study of the psychology of hearing, we should know something of the principles of acoustics, and also something of the anatomy of the ear. Accordingly in our first two chapters we shall review the physical basis of sound, and describe the organ of hearing.

[1] For more detailed accounts of this subject see the following numbers in the list of references: *4, 9, 17, 18, 35, 144, 148.*

GENERATORS AND RESONATORS

All generators of sound are vibrating bodies. In order
that the generator shall be an effective source of sound, it
must vibrate with a certain frequency. Roughly considered,
frequencies between the limits of 16 vibrations and 20,000
vibrations per second are the natural sources or stimuli of
sound. But in order that the sound may be heard, two further conditions must be met. First, an appropriate medium
must communicate the vibrations of the generator to the
ear. This medium is usually the air, though other gases,
as well as liquids and solids, may serve a like purpose. Most
students of elementary physics are familiar with the experiment in which an electric bell is placed under the glass jar
of a vacuum-pump. One sees the bell through the glass,
and hears its ringing. Then, as the pump is set in operation,
and the air within the jar gradually removed, the sound
grows fainter and fainter, until at last the vibration is no
longer audible. One continues to see the clapper vibrating
against the bell as actively as ever, but no sound is heard.
The exhaustion of the air does not interfere with our sight
of the bell, yet its removal reduces, and finally eliminates,
all sound. This suggests that some material medium is
requisite for the propagation of sound-waves.

In addition to an appropriate medium, a certain intensity
of the sound-wave is requisite for hearing. The vibration,
if too weak when it reaches the ear, remains inaudible. Although the intensity of the sound is partly determined by

the force with which the generator sets the air in motion, this force, as a rule, is greatly magnified by the resonance of other bodies adjacent to the generator. Thus, the sound *a* of a musical instrument is usually amplified by means of a resonator which is attached to the generator. With stringed instruments, for instance, it is the strings that initiate the sound, but it is the body of air contained in the sound-box behind the strings that, when set in motion, intensifies the air-waves sufficiently to make them audible. The flaring bell of horns in wind-instruments serves a similar purpose. But even surrounding objects, such as the floors and walls *b* of rooms, may respond to the action of the generator, and serve as resonators to intensify its sound. Thus the conditions under which a sound is propagated within a building are very different from those under which sound is carried in the open air. Surrounding objects not only resonate and intensify sounds; they also reflect sounds—as is illustrated *c* by the echo, which takes place under conditions that permit a direct reflection of the sound to its generator after a period of time determined by the distance of the reflector.

REVERBERATION

Reverberations set up. by the reflection of sound in a closed space are of special importance in the architectural design and construction of auditoriums. An empty room sounds "hollow," and its reverberations may be so confusing to the hearer that neither speech nor music can be readily understood. To secure appropriate acoustics in an auditorium is a problem the solution of which demands great foresight and care on the part of the builder. The size and shape of an enclosed space, and the material of which its walls are made, all are determining factors in the reflection, absorption, and resonance of sound. With respect to the materials of construction, density, porosity, and elasticity are involved. F. R. Watson (*130*) finds that sounds are reflected in proportion to the density of the material of which a wall is constructed. Again, if the wall is smooth and even, reflection takes place just as a beam of light is reflected from a plane mirror, whereas if the surface is rough or irregular in contour, the waves will be distributed like the diffused reflections of light. Porous bodies will transmit sounds in much the same proportion as they transmit air. Thus porous walls will absorb sound according to the degree of their porosity. If the walls of a room are elastic, they will resonate, and thus intensify sound; they will also reflect or absorb sounds that are not in tune with their various periods of resonance, in accordance with their density and porosity.

Thus it is easy to see that reverberation and its correction offer no simple problem, and that to obtain acoustic properties that are appropriate to the uses of an enclosed space such as an auditorium is not always easy. The disturbance occasioned by reverberations can often be corrected by covering the walls of a room with fabrics or felt which will absorb the sound; but to secure the necessary reflection and resonance of sound, so that a voice will carry well and be easily heard in all parts of an auditorium, is more difficult; for good acoustics depend both upon the materials of reflecting surfaces and upon the shape and size of the space in which the sound-waves are distributed.

The velocity with which a sound is propagated through the air is about 340 meters per second at a temperature of 16 degrees centigrade With increase of temperature the velocity increases, though it is independent of air-pressure. When different strata of the air are differently heated, a refraction of the sound-waves takes place (*144*, 10 f.). For instance, on a hot, still day the layer of air in contact with the earth becomes heated, and expands. As a result the sound travels more rapidly near the ground than in the layers above, and consequently the waves are drawn upwards. For this reason the sound does not carry well into the distance. In the evening of such a day the earth radiates its heat quickly, and so cools. The first layer of air to cool is the one adjacent to the earth, which becomes more dense. The effect now is to draw down the more rapidly-moving sound-waves from above. Thus the sound carries farther than it otherwise would, an effect which is especially noticeable if the sound is moving over a smooth surface, such as a sheet of water.

SOUND-WAVES

Although our primary interest is rather in sounds as they are heard than in the details of their physical being, we must know something of the nature of sound-waves in order to understand the physical conditions of hearing, and thus be able to interpret psychologically what it is that we hear.

Sound-waves are phenomena of vibration, and hence they require an elastic medium, the particles of which can swing freely back and forth. As these particles oscillate they set in motion other particles with which they come in contact; accordingly, the sound is dispersed in all directions. Certain particles are more resistant than others, owing to the varied conditions of their respective media. Thus some substances, hard and smooth, have the effect of reflecting the waves; others, soft and penetrable, absorb them; while still others, being elastic, are readily set in sympathetic motion, and act as resonators of the sound.

The characteristic vibration of an elastic medium, since it consists in a to-and-fro movement of its particles, is a pendular motion. With reference to the number of vibrations back and forth in a given period of time, we speak of the *vibrational frequency* of the sound. This frequency is usually measured by the number of complete oscillations per unit of time in which a particle swings, like the bob of a pendulum, first in one direction, then in the other, and finally back to its initial position. When we say that the ear detects sounds within the range of 16 to 20,000 vibrations per second, we are referring to complete vibrations, back and forth, which we shall designate hereafter as "v.d." (*vibration double*). This notation is here specified because the half-vibration period is sometimes made the unit, especially by French scientists.

While the rate of vibration is an essential condition for the kind of sound we hear, determining as it does the factor we describe as *pitch,* there is another important characteristic of the sound-wave, its *amplitude.* It is a known fact that a pendulum of fixed length will swing with a definite frequency without regard to the amplitude, provided the amplitude is small. With this limitation, the greater the amplitude, the greater will be the average velocity; accordingly, the number of the oscillations per unit of time will remain constant. Likewise when a stretched string generates a sound, it may be plucked lightly or heavily, and the period of its vibration will remain constant, although the amplitude of its movement greatly varies. This change of amplitude occasions the differences in loudness of the sound heard. Amplitude of vibration is therefore the essential condition of *intensity.*

A third feature of the simple pendular-formed sound-wave is its *phase.* Since a particle of air in motion has a to-and-fro excursion, its position is constantly changing. Its phase can therefore be defined as the particular position it occupies with reference to its normal or original place of rest. If we plot the various positions which the particle must take in a complete vibration by expanding it into a curve, we obtain the accompanying figure (Fig. 1), in which the horizontal line represents the place of rest, while the curves above and below the line represent the successive instants in which the particle moves out of its original position to and fro like a pendulum. Indeed, this curve can be described by the bob of a pendulum provided with a marker, if the paper on which the marker registers its position is moved evenly at right angles to the path of the pendulum. Similar curves are often obtained for purposes of registration and measurement by attaching a marker to one prong of a tuning-fork set in vibration, while the surface upon which the marker records its course is moved gradually

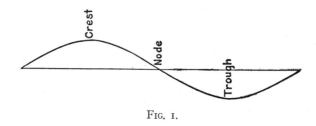

FIG. 1.

and continuously at right angles to the path of the tuning-fork's prong. In plotting curves of this kind one refers to *nodes, crests,* and *troughs,* the nodes being the places of rest, or the points where the curve crosses its axis, while the crests and troughs are the points which indicate the greatest excursion of the moving object above and below this axis.

Attention has recently been called by Otto Abraham (*2, 3*) to a fourth feature of the sound-wave, which under special conditions may manifest itself in consciousness. In a simple pendular vibration the wave-length is a function of the vibrational frequency; so that the two terms are used interchangeably. But under special conditions

Abraham seems to have demonstrated that the wave-length
may be modified without altering the vibrational frequency.
These conditions are provided in the sounds produced by
a siren. The siren is a circular metal disk punched with a
series of holes at regular intervals equidistant from the
centre. The disk is then rotated, and a jet of air focused
upon the series of holes. The regular frequency of pulsations that ensues gives rise to a tone varying in pitch with
the speed of rotation. As the jet of air passes through the
succeeding holes in the disk, sound-waves are propagated
in the air, the crest of each wave occurring synchronously
with each pulsation. During the moments in which the
jet of air is imprisoned by the closed surface of the disk,
the elastic air-particles swing back to form the trough of
the wave. The total length of a single wave is measured
from crest to crest, or, roughly, from pulsation to pulsation,
and so long as the distance between the holes of the disk
remains constant, and the speed of rotation uniform, the
sound will have the same character. But it is quite possible
to introduce the same number of holes in circles at different
distances from the centre. With a uniform rotation the
same frequency of pulsation will then be given by each of
these circles, but the relation of crest to trough will vary;
because near the centre the holes must be closer together,
and the periods corresponding to the troughs of the wave
shorter, whereas near the periphery the time in which the
crest is being formed must be relatively shorter, and the
period between pulsations longer, as is suggested by the

accompanying figure (Fig. 2). Abraham's observers report that this difference is phenomenally apparent when one compares siren-tones of like pitch produced by pulsations of air in which the time-relation of crest and trough is varied. When the trough is lengthened a *brighter* tone is produced, and when it is shortened a *duller* tone results, although the pitch, which corresponds to the frequency, remains the same.

Frequency and amplitude, however, are the two chief features of the physical sound-wave. Still our consideration of the physics of sound is not complete with this description,

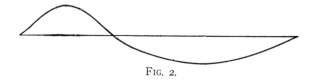

Fig. 2.

for when we recall the fact that sounds depend upon generators, resonators, and sound-absorbing and sound-reflecting bodies, we may at once suspect that the sound-waves actually conducted to the ear are seldom simple or constant, as regards either their frequency or their amplitude of vibration. Indeed, they are just the reverse. Usually the ear is beset with a mass of vibrations, the resultant of numerous frequencies and correspondingly numerous amplitudes. To this mass of sound the ear must make its adjustment, and from the sound, as heard, must derive such information as will make hearing mentally useful. We shall in due course refer to the parts played by the ear and by the mind in effecting this adjustment for purposes of analysis and correlation; at present, our interest being the sound-wave itself, it is important to note that these waves are usually complex, being made up of a number of different frequencies, and a number of different amplitudes of vibration.

PARTIAL VIBRATION

As regards complexity in the frequencies of vibration, our knowledge may be extended by a study of the partial vibrations set up in the sounding body, the generator itself. If one experiments with a stretched string, or monochord, one finds, upon bowing it or plucking it at different points along its length, that while the general effect of pitch remains the same, the "quality" of the sound varies. If we examine the vibration of the string under these conditions, we can give an explanation for this alteration in the sound. When the string is bowed in the middle, the excursion is greatest at this point, and gradually diminishes towards the two fixed ends. But when the string is bowed at a point one-fourth its length from one end, the resultant vibration has two characteristic forms. One, as before, will be the vibration of the whole string. The other, included within the total vibration, is the excursion at the point of bowing, which is now greater than it was in the first instance. As a result, we have two tones instead of one. The string is set in motion as a whole, and produces its characteristic sound. This is called the *fundamental*. But there is also a weaker tone added whose frequency of vibration corresponds to that of half the string (Fig. 3).

Thus, while the string's length, tension, and mass per unit of length determine the fundamental, the location of the bowing determines a subsidiary tone, as indicated in the figure. Indeed, it usually determines several such tones; for, generally speaking, when the string is once set in motion, it vibrates in any and all ways possible to it. Therefore, in

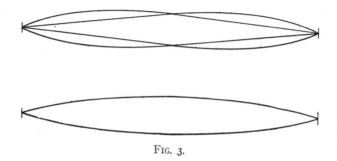

Fig. 3.

addition to the fundamental vibration, there occur numerous subsidiary vibrations which condition a whole series of *overtones*. The chief restriction upon these partial vibrations is that the string shall divide itself into equal parts. Such divisions are favored because the string is swinging as a whole, and symmetrical divisions in some measure obviate the interference of one partial vibration with another. Practically, then, a vibrating string vibrates not merely as a whole, but also in halves, thirds, fourths, fifths, sixths, etc. This means that, in addition to the fundamental frequency of n vibrations per unit of time, the sound is augmented by subsidiary vibrations called *overtones,* whose frequencies are $2n$, $3n$, $4n$, $5n$, $6n$, etc. The overtones are normally of lesser amplitude than the fundamental, and correspondingly weaker. As a rule their amplitudes gradually decrease in passage from the lower order of the larger partial vibrations to the higher order of the smaller partials. Yet peculiarities of the instrument and different modes of sounding it lead both to the emphasis of some partial vibrations and to the weakening or destruction of others.

We have selected the string, or monochord, for our illustration, because here a demonstration is so simple. With a set of small paper riders hung over the string at appropriate intervals, the existence of these partial vibrations can readily be made apparent. For instance: place a light rider *a* at a point one-sixth the length of the string from one end, and another *b* at a point one-third the length from the same end. By bowing the string at one-sixth its length from the opposite end, the fundamental will be sounded, while at the same time the partial vibration $3n$ will be intensified. If the manipulation is carefully executed, the little rider *a* will fly off, while the other *b*, though agitated, remains upon the string. The experiment shows that within the wave engaging the string as a whole, a subordinate wave has been set up, which divides the string equally into three parts. Nodes of relative quiescence have been formed at the two divisional points, allowing rider *b* to remain upon the string, while crests of relatively greater excursion have been formed at three other points corresponding to the division into sixths. Having been placed at one of these crests, rider *a* is forced from its seat (Fig. 4).

This, then, is one of the characteristics of vibrating instruments: they tend to generate a complex of partial vibrations, which are constituents of the whole sound, and are found to embrace a series of frequencies expressed by the formula *n*, 2*n*, 3*n*, 4*n*, 5*n*, 6*n*, etc. Both the stringed and the wind-instruments vary greatly in the number and prominence of their partials, and this divergence largely accounts for the different "quality," or "timbre," of tone ob-

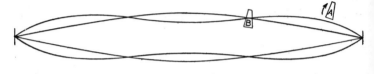

FIG. 4.

tained from different instruments. The violin, the horn, the harp, and the clarinet, each has its characteristic timbre. To illustrate this fact, Professor Dayton C. Miller has kindly supplied me with the accompanying photographs of the tones of four different musical instruments, each sounding a fundamental tone of approximately the same pitch, namely, middle c, 256 v.d. (Fig. 5). The remarkable variation in the complexity of these four curves is occasioned by the presence of different overtones with different degrees of intensity.

With these facts in view, the organ-builder constructs his various stops; for by an appropriate combination of the

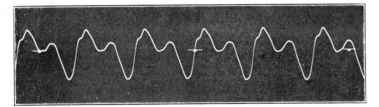

FLUTE. Fundamental frequency 260 v.d.

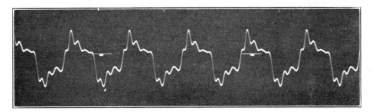

CLARINET. Fundamental frequency 257 v.d.

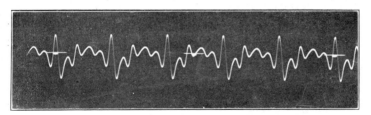

OBOE. Fundamental frequency 248 v.d.

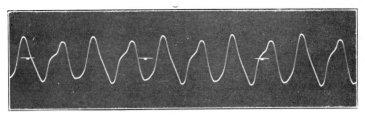

VIOLIN. Fundamental frequency 263 v.d.

FIG. 5.

tones of a series of pipes, due regard being paid to their relative intensities, it is possible to imitate closely the subtle qualities of the human voice and also those of a variety of musical instruments. As we shall see when we come to consider the sounds that correspond to these partial vibrations, the fact of their existence has an important bearing upon the nature of our sense of hearing and upon the use to which we are able to put it.

COMBINATIONAL EFFECTS

Subsidiary tones are not only inherent in a single clang; they are also produced by combining two or more separate sounds. Among these are the *combination-tones*. The propriety of treating these as a normal constituent of the sound-wave is questionable, for they are commonly supposed to have their origin within the ear. Yet "objective" tones of this order do occur, and for other reasons it seems desirable at this point to give them brief notice. Combination-tones are conditioned by the simultaneous sounding of two or more tones of different vibrational frequency. Among the combination-tones are *summation-tones*, which correspond to the frequency of two sounding vibrations when added together; and *difference-tones*, which correspond to the frequency of two sounding vibrations when subtracted. Thus the summation-tone for 200 v.d. and 300 v.d. would be a tone of 500 v.d., while the difference-tone in this case would be a tone of 100 v.d. Difference-tones of other orders than those corresponding to the simple difference of the generators are also observable. In the case of 300 v.d. and 400 v.d., the difference-tones might be those of 100 v.d. and 200 v.d., the last named marking the difference between the first difference-tone and the lower generator. These tones are sometimes important constituents of the sound-mass, and will be referred to again.

Another subsidiary tone is the *mean-tone*. This corresponds to the arithmetical average of the vibrational frequencies of two generators. Thus with 200 v.d. and 250 v.d., simultaneously generated, there may be a mean-tone corresponding to 225 v.d. On the whole, mean-tones have little significance. A greater importance attaches to the fact of *interference* among sound-waves.

INTERFERENCE

We have spoken of resonating bodies that can be set in sympathetic motion so as to augment the sound-wave by repeating its frequency. These bodies are more or less selective in their function. Some resonators having no definite vibrational frequency of their own respond to any tone or combination of tones. Thus the sounding-board of a piano resonates to all the strings, and more or less amplifies all tones of the instrument. Other bodies are more highly resistant to any resonance other than that of their own natural period of vibration, including, of course, any of its partials. A bit of bric-a-brac in a room may be set in vibration by a single string of the piano so as to occasion a disturbing effect on account of the unusual timbre which accompanies the tone of the vibrating bric-a-brac. This is but a special case of resonance. Ordinarily when a sound-wave meets objects that do not have its own period of vibration, the shock is either deadened and absorbed, or reflected without being augmented.

Whenever two generators are in motion together, interference is likely to result. If the two generators have the same frequency, they may either augment or decrease each other's intensity. When they vibrate together in the same direction, or *phase*, the resultant sound is intensified. When they move in opposite directions, there will be interference, and they may even nullify each other. This can easily be demonstrated by a simple experiment. Let a tuning-fork of fairly short wave-length, say 600 v.d. to 700 v.d., be sounded at a suitable distance from a good reflecting surface such as the wall of a room. If we vary the position of the fork, and listen with a single ear in the intervening space between the fork and the wall, it is possible to find one or more places where the sound no longer can be heard. At these points the oncoming wave from the generator meets with sufficient interference from the reflected wave to reduce its amplitude below the threshold of audibility, whereas an amplification of the sound will accompany any change in the position of the ear, either towards the wall or towards the fork.

BEATS

If there is a slight difference in the vibrational frequencies of two simultaneously acting generators, there occur an alternate increase and decrease in the intensity of the sound. The tone seems to oscillate, and these oscillations are known as *beats*. When we plot the curves of these sound-waves to show the exact nature of this interference, we find it to be a matter of phase-difference, one generator exciting vibration in one direction while the other is exciting an opposite motion

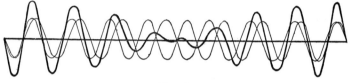

F<small>IG</small>. 6.

of the same medium (Fig. 6). It also appears that these interferences occur at set periods of time, so that the difference in vibrational frequency also measures the number of beats in each second. Thus, if a tone of 256 v.d. is sounded together with one of 260 v.d., there will be four beats per second. Four times in this period the resultant sound has been weakened and intensified. We hear, not a smooth regular tone, but a tone which pulsates or beats.

Since beats correspond to the difference in vibrational frequency of the two generators, it is easy to infer that difference-tones are merely a result of beats of high frequency. This, indeed, is the view taken by many physicists, who re-

fer to difference-tones as "beat-tones." The inference, however, is of doubtful validity, because it has seemed impossible
to detect the vibrational existence of many quite audible difference-tones in the resonating medium of the air. Specially
adapted resonators do not, as in the case of overtones, reveal
the objective presence of difference-tones; yet the existence
of these in the phenomenal sound-mass can not be doubted,
and these tones again may be made to beat with objective
tones that are near to them in vibrational frequency. For
this reason their origin is usually ascribed to a function of
the ear, rather than to a vibration of the air. Such arguments have led to the consideration of combination-tones as
physiological rather than physical phenomena of sound.

Beats are readily heard as fluctuations up to about 30 in
the second. Beyond this we note a certain "roughness" of
the tonal mass, which finally vanishes, according to Helmholtz (*35*, 171),[1] when the difference in vibrational frequency has reached the limit of about 130 per second.
Stumpf (*122*, II, 461), however, remarks that roughness
may be detected in higher regions of the scale when the
beats are as frequent as 400 in the second.

We have seen that our stimuli of sound are for the most
part complex masses of air-waves, consisting of varying frequencies of vibration with varying amplitudes. To a certain extent they may be analyzed into components of regular frequency and constant amplitude; but not altogether
so. Many waves change their frequency by incidence with

---

[1] Roman numerals within parenthesis indicate page references.

media of a different resonating character. Intensity is also subject to manifold changes. It is to irregularities both in the single wave-length and in the more complex masses of sound that we owe many characteristics of the sounds of daily experience. The facts of resonance and interference, however, remain the chief physical determinants of the varied phenomena of sound as we hear it.

CHAPTER II

THE ORGAN OF HEARING [1]

THE GROSS STRUCTURE OF THE EAR

THE ear is a physiological mechanism nicely adjusted both to receive and to analyze the complex vibrations of the air that strike against its outer membrane, the ear-drum. The end-organs, or receptors, of hearing are imbedded in the bony structure of the skull. They are six in number, three being situated on either side of the head. All are labyrinthine cavities filled with watery fluids. It is not certain that the *vestibule* and the adjoining *semicircular canals* may strictly be called sense-organs, though they are capable of stimulation, and give rise to bodily reactions; but the third organ, the *cochlea*, is commonly accepted as the seat of hearing. The vestibule and canals, with their special nerve-terminals, serve the purpose of maintaining the body in its normal upright position, and are therefore known as organs of equilibrium. Two sorts of mechanism are here involved. The first, situated in the vestibule (a small and somewhat spherical cavity, comprising two sacs, the *saccule* and the *utricle*), is stimulated by means of particles of a

[1] For more detailed accounts of this subject see the following numbers in the list of references: *71, 99, 101.*

23

calcareous substance resting upon hairs. As the head is moved these hairs are bent, thus exciting nerves that terminate in the cells from which the hairs issue. This excitation leads to the contraction of certain muscles in the body which serve to correct a tendency to topple over. The vestibule seems to be the most primitive organ of the ear, for we find analogous structures far down in the scale of animal life. Whether the "shake-organ," as it is sometimes called, is ever, properly speaking, a receptor of hearing is uncertain. It has often been so considered, especially with reference to the production of *noise*, but its known function is only that of innervating appropriate muscles to withstand a disturbance of equilibrium.

The second organ of the ear comprises three canals, semicircular in shape, and all opening into a common cavity. These canals are set approximately in the three geometrical planes. Their nerves again terminate in cells bearing fine hair-processes which are to be found in three separate *ampullæ*, or enlargements, one in each canal. The free hairs extending into the water of the canals are brushed back and forth as the water lags through inertia, or is set in motion according to the movements made by the head. In this way more delicate and precise muscular reflexes are introduced to aid the organism in maintaining its balanced upright position. A special sense of equilibrium is sometimes attributed to the vestibule and semicircular canals, but no agreement has been reached on this point because of the difficulty in isolating and distinguishing any unique sensory quality in the complicated experience which always attends a disturbance of equilibrium.

The specific function of these organs as reflex-mechanisms of equilibrium is, however, evident. We need but mention instances in which the organs are pathologically defective. Under these conditions, the individual, when deprived of his sight, finds great difficulty in maintaining his equilibrium. It has been especially noted that deaf-mutes are frequently unable to keep their bearings in the dark, or when swimming under water, post-mortem examination has shown that serious defects of the cochlea are often associated with corresponding defects of the vestibule and semicircular canals. A further indication that these organs are primarily concerned with equilibration is the relatively high development of similar structures in birds and fishes. It is obvious that both these genera should find such a mechanism of especial utility, because they move more freely in various planes than do creatures that tread the earth.

As to the cochlea, we know little of its genetic antecedents, for wherever it has been found, even in fossil remains, it appears as a definite type of structure (27, 135).

With the aid of the accompanying diagram we may note the chief parts of the organ of hearing, and understand something of their respective functions (Fig. 7).

The outer ear, consisting of the *auricle*, or ear-flap, and the external tube, or *meatus*, serves to collect the air-waves. In some animals the auricle has a positive function, in as much as the organism has control of its ear-muscles, and can "prick up its ears" so as to direct the opening towards the source of sound. Man has lost this capacity to so large an extent that his ear-flaps serve only to render a sound coming from the front slightly more intense than a similar sound from behind.

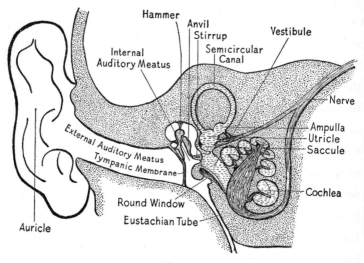

FIG. 7.

The external auditory meatus is closed by the *tympanic membrane* of the ear-drum. This membrane contains fibres radiating from a centre that is drawn inwards by its attachment to a small bone supported by a muscle, the *tensor tympani*. The drum-membrane is thus held taut, so that it responds to agitations of the air by duplicating in transverse movement the complicated form of the air-wave. Behind the membrane is an air-filled chamber known as the middle ear, or *internal auditory meatus,* a cavity connected by the *Eustachian tube* with the air-passages of the nose. The Eustachian tube, ordinarily closed, is opened in the act of swallowing. At such moments we can clearly detect a pressure against the ear-drum from within. The tube serves as a safeguard to our hearing by helping to maintain a constant air-pressure on either side of the membrane. When a catarrhal affection stops the openings of the Eustachian tubes, we sometimes experience a temporary deafness, by reason of a difference in the density of atmosphere on the two sides of the drum-membranes. If an undue tension is thus produced, the drum fails to respond with the delicacy normally required of it. A relatively sudden change in air-pressure can be noted when we descend by elevator from the higher stories of a "sky-scraper." Pressure on the ear-drums is felt as we pass downwards into the denser strata of air, but this pressure can be quickly relieved by an act of swallowing; for in opening the Eustachian tubes we at once equalize the pressure on either side of the membrane.

We have mentioned the small bone whose process lies along the inner surface of the drum-membrane, terminating at the centre. This bone, known as the *hammer*, articulates with another small bone, the *anvil*, which in turn articulates with a third called the *stirrup*. These three bones, all situate in the middle ear, are the ones referred to in our books on elementary human physiology as the "smallest bones of the body." According to a commonly accepted view, these bones serve the purpose of communicating the movements of the drum-membrane to the inner ear where the nerve-terminals of hearing are situated. But not only do they serve as a bond of communication between the outer and the inner ear; they also act as a system of levers to increase the force, while at the same time decreasing the excursion, of the vibrating movement communicated by the air. The importance of this mechanism is evident when we reflect that the stirrup pushes against another membrane, beyond which is a column of watery fluid. Since the liquid is not elastic, it must be displaced in order that the stirrup may move at all. For this an increase in force is necessary. The actual displacement of the stirrup is still effective, even though its excursion is far less than the corresponding movement of the elastic drum-membrane.

The small oval plate of the stirrup is set against a membrane-covered window opening into the labyrinth. This opening is known as the *oval window*. As in the case of the semicircular canals, the cochlea is filled with a watery fluid which is only to a slight degree compressible. In order that the stirrup may move at all, there must be some outlet for the fluid against which it presses. Such a release is effected by another opening, also covered with an elastic membrane, and known as the *round window*.

We have now before us the gross features of the ear's structure. The air-waves that impinge against the drum-membrane set it in sympathetic motion. This motion duplicates in a transverse direction the form and amplitude of the sound-waves, in much the same manner as does the disk of a telephone-receiver. The motion is then conducted without serious modification to the stirrup, which acts somewhat like a piston against the watery fluid of the inner ear; not exactly, however, for it seems to be immovable at its base, so that it swings like a door on its hinges (*35*, 134). A movement duplicating the physical sound that has stimulated the external receptor is thus set up in the cochlear liquid.

A somewhat different view of the action and function of the chain of bones which lie between the drum and the cochlea has been advanced by Gustav Zimmermann, an ear-specialist in Dresden (*149*). Zimmermann contends that this mechanism is protective and accommodatory to the action of the watery fluids of the cochlea, instead of being an essential means of conducting sound to the inner ear. His views, although they have not met with general acceptance, are worthy of note, because certain pathological phenomena seem to indicate that sounds are transmitted even when the bones of the ear fail to function. According to Zimmermann, sound is conducted chiefly through the middle-ear cavity to the bony *promontorium* which is situated opposite the drum-membrane. The vibrations of this elastic plate of bone are thus directly communicated to the watery fluids of the cochlea which lie within this bony wall. Consequently, the function of the stirrup and its leverage is not so much to communicate the movements of the drum-membrane with a reduced amplitude, as it is to regulate excessive movements of the watery fluid, particularly those of great amplitude occasioned by sounds of slow frequency, and hence, defects in the operation of this mechanism are associated with deafness to low tones, and subjective noises. It is not improbable that the bones of the ear serve both these functions, and that sound is communicated to the inner ear through the stirrup, through the promontorium, and also through the round window.[1]

Pohlman (*89*) has more recently come forward with a theory of the mechanics of the middle ear, which, while it attributes to the ossicles the function of transmission, regards them as a rod rather than as a set of levers in their action.[2] He argues that there can be no inertia in this apparatus, because, if there were, it would be impossible to communicate the ultra-microscopic excursions of the drum-membrane in its response to ordinary sounds. His conclusion is that the sound-pulse must be directly transmitted to the inner ear, thus providing for a direct action upon the nerve-terminals, as does the theory of Zimmermann; with this difference, however, that Pohlman's anatomical investigations lead him to believe that the ossicles are not a means of regulating perilymphatic pressure, but are the chief instrument in the communication of the sound-pulse to the inner ear.

THE AUDITORY NERVE-TERMINALS

We have stated that the auditory nerves terminate in the inner ear. We must now consider the nature of their distribution, and the character of the structures surrounding

[1] Cf. Beyer (*12*). [2] Cf. also Schulze (*103*).

them. This will acquaint us with the manner in which the nerves receive the stimulation from without, and also give us some insight into our ability to discriminate the complex sounds we hear. So far, our account has indicated only that the sound-waves arising from without the body are transferred in all their complexity to the fluids of the inner ear.

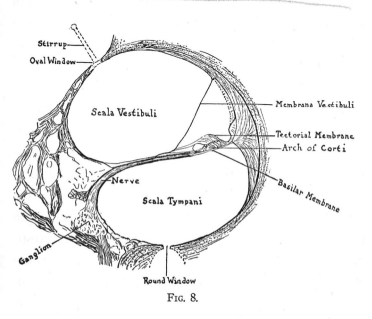

FIG. 8.

The particular fluid concerned with the sense of hearing is confined in the spiral tube called the cochlea. This tube is coiled like a snail-shell two and one-half times before it reaches its apex. It is divided through the middle by a partition, half of which consists of a bony shelf, and the remainder of a complicated formation of tissues; within this tissue the nerves of hearing terminate (Fig. 8). Running in beneath the bony shelf, they distribute themselves all along the partition of tissue from the base to the apex of the cochlea; in this way the cochlear canal is divided into two chambers, the upper of which is known as the *scala vestibuli,* and the lower as the *scala tympani.* Against the base of the upper chamber the stirrup acts as a hinged piston, while at the base of the lower chamber we find the round window. At the apex of the cochlea there is an opening, the *helicotrema,* which gives communication between the two chambers.

Should we ask the question, how the column of fluid is displaced when it is set in vibration, we find two possible answers. If the partition between the two chambers were quite rigid, the fluid would be displaced as a single compact body at the round window; this would be possible because of the opening at the apex of the cochlea. But since the partition is half-membranous, it must itself be subject to some compression or "give." Consequently, the pressure exerted against the round window may be communicated, in part or in whole, through these tissues, as well as by way of the opening at the apex.

Let us now direct our attention to the tissues in which the nerve-terminals are embedded. In the first place the entire structure is set off and protected by a thin membrane, the *membrana vestibuli,* that arises at the edge of the bony shelf, and extends above the structure of tissues to the outer wall of the cochlea. This canal, the so-called *scala media* of the cochlea, contains *endolymph,* and opens into the saccule, while the *scala vestibuli* and the *scala tympani* contain *perilymph,* and have no opening into the other canals of the labyrinth. Another structure known as the *tectorial membrane* also arises from the bony shelf, and stretches over a portion of the floor of the *scala media,* but its outer lip is suspended in the *endolymph,* and does not reach the cochlear wall. As for the floor of this canal in which the nerves terminate, its foundation is a series of cross-fibres closely bound together, and extending from the bony shelf to the opposite wall. This is the *basilar membrane.* Above it enter the nerves, terminating in cells erected upon the basilar membrane, and provided with little brush-like hair-processes that project into the tectorial membrane. The nerve-cells are supported by a cellular tissue, and are held in angular position by an arched structure known as the *arch of Corti.* In this way the brushes are brought approximately at right angles into contact with the surface of the tectorial membrane; and according to the observations of Keith (*145,* 210) and Wittmaack (*141*) they are embedded in its tissue. It seems probable that one factor, at least, in the stimulation of the nerves by a commotion of the endolymph is the contacts that would be made between the tectorial membrane and the little hair-processes of the nerve-cells.

Among all the parts of this somewhat complicated structure, the one that has received most attention as the probable seat of hearing is the basilar membrane. Consisting of a series of transverse fibres, estimated as some 10,000 in number, it extends virtually throughout the length of the cochlea from base to apex. But whereas the cochlear tube grows smaller as it approaches its apex, the transverse fibres are relatively short at the base (0.17 mm.), and gradually increase (to approximately 0.5 mm.) in length as the membrane extends throughout the coils to the tip of the tube.

We shall now examine several hypotheses that have been offered to explain the physiological action of the basilar membrane when it is stimulated in the production of sound. This examination will bring the receptors of sound into closer view, and will also serve to define the problem that confronts the physiologist. We may state at the outset that all the fruitful theories of sound-reception have been founded upon two conceptions: One is the conception of *resonance,* and is based upon the analogy of the harp, an instrument of many strings, each attuned to its special quality of sound. The other renounces the principle of *resonance* as being inappropriate to the action of the basilar fibres, and instead makes use of the principle of *displacement.* The first of these explanations assumes an elaborate set of resonating strings, while the second relies upon the mechanical action of forced movements in a non-resonant membrane. A third view mediates in some ways between these two conceptions and regards the action of the basilar membrane as analogous to that of a telephone-receiver, which duplicates the complex forms of sound-waves by sympathetic vibrations of its disk.

THE THEORY OF HELMHOLTZ

The best-known of the physiological theories of hearing is the "harp"-theory associated with the name of the eminent German physicist, Hermann von Helmholtz. According to Helmholtz (*35*, 143 ff.) each transverse fibre of the basilar membrane has its own vibrational frequency, and therefore resonates whenever this particular frequency happens to be a component of any vibration that is set up in the cochlear fluid. We have here a simple explanation for the capacity of the ear to analyze the resultants of air-waves of varying frequencies and amplitudes. Since the short fibres are situated near the base of the cochlea, these are responsible for the higher sounds, while the longer fibres at the apex are the special organs of the lower register of tones. Each fibre is supposed to stimulate a single nerve-terminal, adapted by means of a "specific energy" to the frequency of the fibre. When the appropriate nervous discharge reaches the brain, we hear the corresponding tone as a more or less discrete unit, or "element," of consciousness.

Although the idea basic to this interpretation has been subjected to many minor modifications, both by Helmholtz himself and by later investigators, it to-day remains in its essential characteristics the prevailing theory of the reception of sound. The fact that we are able to analyze numerous components of a complex sound, such as the different partials in a clang, has contributed to the spread of the theory, because it affords so simple an understanding of the ear's analytic capacity. The pattern or form of the external sound-wave is repeated with all its varied frequencies, amplitudes, and interferences, in the movements of the cochlear fluid; and the large number of differential resonators, which the theory assumes, are regarded an appropriate means of picking out these increments of the total sound-mass, and communicating them severally along different nerve-paths to the brain. The resultant analysis in consciousness is made plausible if we suppose that each of these nervous components has contributed its individual conscious element, which attention brings separately into prominence whenever a psychological analysis is undertaken.

Further evidence in support of this theory is furnished by certain cases of abnormal hearing. These are cases in which "gaps" and "islands" are detected in the individual's consciousness of sound. That is to say, a person may be deaf to all the sounds of a certain region in the serial progression from low to high tones. The existence of more than one gap may occasion islands of sound set off by regions of deafness below and above it in pitch. Such cases are readily explained if we assume a pathological condition of the fibres that correspond to the regions of the tonal gaps. More definite evidence bearing on this point has been obtained by experimentation with guinea-pigs. The animals were first subjected to prolonged stimulation with very intense sounds of a certain pitch, and then killed. Post-mortem examination showed that the delicate structure of the basilar membrane had been injured in those regions of the cochlea which are regarded, according to the theory of Helmholtz, as the probable receptors of these particular sounds (*139, 147*). Other experiments and observations likewise point to the conclusion that low sounds are produced at the apex and high sounds at the base of the cochlea.

THE THEORY OF MEYER

In opposition to the Helmholtzian view is the displacement-theory formulated by Max F. Meyer (*64*). Meyer contends that the analysis of sound is not made on the basis of resonance, but rather from contacts or impulses of the nerve-endings. So far as a particular nerve is concerned, it is not specifically adapted to one frequency alone, but may respond to a large number of different frequencies. Each frequency of the nerve-impulse is correlated with a certain sound in consciousness, and one nerve alone might be capable of producing the whole gamut of simple sounds. This principle of interpretation is sometimes extended to embrace a "telephone-theory" of hearing. The basilar membrane of the cochlea is supposed to reproduce, sympathetically, the motion of the fluids so that the nerve-endings distributed throughout the membranous tissue will receive shocks of different frequencies and intensities according to their situation, as does the disk of a telephone. The precise nature of the movements made by the tissues in which the nerves terminate, however, has not been agreed upon by different investigators. We shall consider Meyer's view first because his theory does not localize high and low tones, respectively, at the base and apex of the cochlea. Some of the adherents of a "telephone-theory" do accept this provision, which, as already mentioned, appears to be supported by several results of experimentation.

Meyer maintains that the basilar membrane is not an elastic tissue. When subjected to pressure by the action of waves set up in the watery fluids, it functions rather as an indifferent mass. Suppose that the stirrup is communicating a simple air-wave of 200 v.d.; this would cause it to act piston-wise at the rate of 200 impulses per second. The displacement of the column of fluid, against which the piston presses, is made possible by the bulging of the round window. The membranous parts nearest the two windows, that is, at the base of the cochlea, are the first to give way to this pressure; accordingly, they are set in motion. This agitation of a part of the tissues in which the nerve-endings are situated occurs at the same frequency as that of the sound-wave; and hence all the nerve-endings in the portion affected are subjected to 200 shocks a second. Meyer proceeds to explain that, the greater the intensity of the stimulus, the greater must be the excursion of the stirrup, and the longer the stretch of tissue set in motion. Since the number of nerve-terminals affected depends upon the stretch of tissues involved, he concludes that variation of the intensity of the sound as heard is correlated with the number of nerve-terminals stimulated. Ordinarily the resultant of a series of complex vibrations, rather than a single frequency, is operative, but it·is possible to calculate from the form of the wave the different sections of the membranous partition that would be involved, and the different frequency with which each section would respond. The analysis indicates that the various constituents of the sound-mass, both its partials and its combined fundamentals, will affect different sections so as to provide for the separate hearing of its several components. Meyer also finds, and this is perhaps the special merit of his theory, that under certain conditions additional tones, not present in the air-wave, will be produced. These are the so-called "subjective" combination-tones to which reference has already been made (cf., p. 17).

We have remarked that difference-tones seem to be produced under conditions that exclude the possibility of corresponding components in the air-waves. Helmholtz, who took this view, made various attempts to account for it (*35, 158*). Since the resonance-theory only provides for hearing those components of sound already present in the waves of the endolymph, Helmholtz suggested that subjective combination-tones might be aroused by certain asymmetrical functionings of the ear-drum and the bones of the middle ear. His explanation is not altogether satisfactory; for, as determined by the careful experiments of Stumpf (*116*) and others, the subjective combination-tones constitute a rather extended list, embracing well-defined differences and summations of the producing tones, and no correlation of these combinations with specific asymmetries of the middle-ear mechanism has yet been established. Furthermore, combination-tones may be audible to individuals in whom the apparatus of the middle ear is seriously defective (*100*). Much impressed by these facts, Meyer shapes his theory so as to account for the subjective tones by a specific functioning of the inner ear.

Upon the whole, it is at present inadvisable to take any definite position on this question. We have already observed that physicists are disinclined to accept difference-tones as of a "subjective" order. The most prominent of these tones, the simple difference-tone, is often referred to as a "beat-tone," and is thought to be produced by the rapid beating of the two generators. There is also some evidence (*93, 142*) for the objective existence of both difference- and summation-tones in cases where objectivity has not usually been suspected. It is therefore possible that a special explanation for the production of "subjective" combination-tones may be unnecessary.

A further attempt to explain subjective tones is included in a third theory of audition to which we shall next direct our attention—a theory long since suggested, but only recently stated with some degree of precision.

THE THEORY OF WRIGHTSON

The idea which Sir Thomas Wrightson began to develop over forty years ago, and which is now published with the critical and experimental support of Dr. Arthur Keith's anatomical research (*145*), offers a notable advance over

Meyer's conception of displacement, in the discovery that a single pendular wave may afford four distinct stimuli to the auditory mechanism. These stimuli correspond to four phases of the wave-form: first at its rise, second at its crest, third at its point of crossing the base-line (its nodal point), and fourth at its trough. With the aid of a mechanical device for plotting compound curves, Wrightson demonstrates by a method of counting that, if we take these four phases into consideration, there occur periodic changes to account, not only for the simple tones, in accordance with Fourier's law, but also for the combination-tones (summation- and difference-tones). It can be shown that when this complication of stimuli is transferred to the internal ear, the basilar membrane responds in like manner, to occasion in the cilia of the nerve-cells separate *transverse* movements corresponding to each of these four phases, with periods of quiescence between them. The movements of these hairs, projecting into the jelly-like substance of the tectorial membrane, furnish a discretely tactual mode of stimulation to the nerves, which may in turn be synthesized or analyzed in the cortex. The theory attempts to cover all the chief points upon which the theory of Helmholtz has relied for its support; the mechanical principles are worked out with a refined nicety by means of Wrightson's expert engineering technique, while Keith's experimental research upon the anatomy of the ear strives to make clear its evolutionary adaptation to the purpose at hand.

Yet against this theory there have been raised a number

of objections, which are in part so serious that we must again suspend judgment until an appropriate modification of Wrightson's views has been effected. In particular, the theory seems to explain too much as regards the combination-tones for which it makes provision, for (by following Wrightson's method of analysis), Hartridge (*31*), has shown the theoretical presence of numerous "subjective" tones, including inharmonics, which we do not hear at all, though the analysis indicates them to be as truly "conditioned" as are the tones we do hear. Boring and Titchener (*16*) have also shown that the summation- and difference-tones of the analysis are each an octave too low, and that the ear should always be able to hear not only the tone stimulated but an additional tone one octave lower.

OTHER THEORIES

Numerous other theories have been advanced to explain the functioning of the ear, but virtually all are based either upon the "harp"-idea of resonators, or upon the idea of displacement and the telephone-receiver. Some anatomists have been impressed with the importance of the tectorial membrane as an essential feature in the mechanism of hearing (*30, 110*). This membrane, in which the hair-processes of the nerve-cells terminate, increases its breadth and thickness as it extends towards the apex of the cochlea. Since its specific gravity is but slightly greater than that of the watery fluid in which it is suspended, it appears to be well-adapted to selective vibration. The portions at the base are thus supposed to swing intensively with the higher frequencies, while those at the apex respond to the lower rates of vibration. This suggests a certain modification of the "telephone-theory." As compared with Meyer's view, that any nerve-terminal may respond indifferently to any frequency of vibration, the selective functioning of the tectorial membrane implies adaptation of the nerves in different regions of the cochlea to a graded series of responses to vibration or shock. Such a modification is also in keeping with the apparent fact of tone-gaps and islands of hearing.

Other modifications of the "telephone-theory" are concerned with the nature of the waves propagated along the membranous tissues from the base to the apex of the cochlea. Various attempts have been made to reconstruct these waves with the aid of models of one sort and another, Ewald (24) being the chief proponent of a theory based upon the "sound-picture" of the membranous partition.

The "sound-picture" is indicated by the fixed waves which appear upon a stretched membrane of rubber when it is agitated by periodic vibrations. Ewald suggests that the nodal lines of such a wave differ in distance and length according to the periodicity of the tone. In the case of noise the waves run over the surface without indicating definite nodes. The distances between the nodes of fixed waves is inversely proportional to the number of vibrations, and the analysis of the sound-picture reveals all the components of the sound-mass, including both the partials of a clang and the difference-tones involved. Ewald finds that the waves separate slightly when the intensity of the sound is increased, a fact which seems to explain the lowering of pitch in a sound when the intensity is augmented. The reason assigned to explain a loss of hearing for low tones after the cochlear membrane is partially destroyed is the necessary employment of a large area in the production of all low tones. But the theory is less adequate in its explanation of the loss of high tones, which, according to Wittmaack (*140*), is the usual effect when the destruction is occasioned solely by sound without the accompaniment of jars or shocks to the organs.

H. J. Watt (*134, 162* f.) has formulated the theory that if the membrane be highly elastic, portions of it will be agitated which correspond in size with the frequencies of vibration. A low tone, he thinks, must involve a large section, while a high tone agitates but a small one, the sections beginning at the base of the cochlea and extending inwards. According to his view, the culminating point in any agitation is always at the middle of the section that responds. The fibre thus most acutely acted upon gives the pitch of the tone, while other fibres set in action at the same time contribute to the tonal mass, with an intensity that corresponds to the degree of agitation, and a volume that corresponds to the total range of the fibres engaged.

CONCLUSION

We need not proceed further into the details of the cochlear response to sound-stimuli. It is evident that all physiological explanations are in large measure hypothetical, in as much as we have so few experimental facts to indicate what the mechanism of hearing must be, and only a limited knowledge of the performance of the organ as it actually takes place. The parts of the inner ear are so very small, and so very delicate in structure, that we can not remove them for direct observation under conditions analogous to those that exist when they function normally *in situ*. We are left with the two main possibilities of a resonating or a mechanical response. We know that the nerve-terminals are sufficient in number to provide for the differences in sensational experience that our analysis makes evident. Furthermore, pathological cases, and the experimental destruction of specific regions by subjection to continuous tonal stimulation, make it highly probable that specific sections of the canal are more or less adapted to certain responses. It is doubtful, however, if the response is absolutely discrete; for Held (*34*) has shown that a single ganglion furnishes fibres to an extended region of the basilar membrane. The research of Head and his associates (*33*) likewise forces us to question the assumption that the specific sensitivity of a single nerve-terminal is directly projected in the cortex to furnish the substratum of a simple sensation. If we must give up the idea of "specific energy" as it applies to a single receptor and its communicating tracts to the brain, and must substitute therefor Head's theory of physiological integration with its complicated sifting and sorting of afferent impulses in the production of a cortical process whose conscious correlate is psychologically simple in quality, then the forced movement of the whole basilar membrane would be a closer parallel to the integrative action of the afferent nerves of the skin, upon which Head bases his conclusions, than would be a discrete resonance of fibres in the basilar membrane. It appears more likely, therefore, that a given terminal responds to any frequency within a certain range, and that response to a given frequency is made by a region of nerve-terminals, rather than by a single receptor. Whether the response is that of a resonator, implying that the stretched fibres of the basilar membrane are set in sympathetic vibration, or whether it is only a mechanical arrangement of forced contacts and shocks corresponding in number to the vibrational frequency, is a question we shall not attempt to decide, although the recent papers of Hartridge (32) bring new critical support to the theory of resonance. But whatever the outcome, it is a satisfaction to know that with such simple analogies as the harp and the telephone we can construct alternate explanations, each of which is fairly adequate, and one of which, in all likelihood, will eventually be established as true.

Simply stated, the facts are that the organ of hearing is a mechanism for conducting the air-wave with its complicated pattern into the watery fluids of the cochlea. The agitation thus set up plays over and about a membrane in which the nerve-endings are distributed from the base to the apex of the cochlear canal. By a further mechanism, the details of which are at present less clear, selected nerve-terminals are stimulated, and these innervations, integrating, no doubt, in a complex manner, provide a pattern of nervous activity that correlates with our consciousness of the sound-mass.

CHAPTER III

TONE

THE SENSATION OF TONE

In dealing with the sensation of sound, we may begin with a general classification of auditory experience by dividing all sounds into three groups: *tones, vocables,* and *noises.* That is to say, every sound we hear can be referred to one of these classes; though the reference is not always simple, and the classification is often in large measure determined by the particular setting of the experience in which the sound chances to be heard.

Generally speaking, however, tones are sounds which have a musical context; we hear them as the sounds of musical instruments, accompanying and succeeding one another in patterns expressive of musical thought. Vocables, in turn, are sounds whose context is linguistic; they suggest the utterance of a speaker, and the communication of ideas and feelings by word of mouth or exclamatory ejaculation. Noises, finally, belong to occurrences in the material world about us; they denote physical change and the movement of objects, such as the disturbances of wind and rain, the falling of bodies, the whir of machinery, and the like. The broad and general significance of noise suggests that it is probably the earliest apprehensible type of sound.

48

Only when we analyze sounds apart from their associative contexts do we come to realize that what we call a perception has as its nucleus a definite configuration of members, the accurate description of which furnishes an analysis of the sound itself. Upon such an analysis we are about to enter. When it has been achieved, we shall find that the same elements that in one configuration give us tone, in another give us the vocable, and in a third, noise. Yet the configurations of each class are so variable that by a gradual transition we may pass from one to another, or we may find one overlying another. All the sounds that exist are perceived as tones, vocables, or noises, according both to their inherent configurations and to the use we make of them. For instance, the sound of the kettle-drum, which plays an important rôle as a musical instrument, may also be heard under conditions that suggest now a rumbling noise, or again the murmur of human voices.

Our classification, therefore, primarily serves as a guide to the investigation of sound. Yet the guidance we derive is sure; for we find each class possessing a typical configuration which acts upon us in a compulsory manner, so that when the conditions of hearing a particular pattern are adequate we must perforce hear a tone, a vocable, or a noise, according to the nature of the sound that affects us. Confusion or uncertainty enters in only when the pattern is obscure or the context shifting.

How then shall we proceed to the analysis? The most natural approach is from the side of physics. We have already noted that sounds are occasioned by vibrations of an elastic medium, and the physicist tells us that sound-waves are constructed of simple pendular-formed components. But physical simplicity and psychological simplicity are by no means the same; nor does physical complexity of stimulation necessarily condition a corresponding complexity of sensation. In vision a neutral gray of perfectly simple quality is occasioned by a mixture of varied frequencies of light-wave, a physically complex structure, while an intermediate color such as orange, which, since it resembles both red and yellow is qualitatively less "simple" than either, may have as its physical correlate a homogeneous, or physically uniform, light-wave. Nevertheless, recourse to the physical stimulus is necessary; for it is one of the conditions that must be brought under control before a psychological analysis can take place. Let us, then, look into the physical nature of tones.

Simple pendular-formed vibrations within the audible range give rise to an integrated unit of sound which can not be analyzed into separable components. We call this experience a sensation of *tone*, although simple vibrations are not the only means whereby a tone can be produced, and the term itself is equivocal, since it is ordinarily used to denote not an "element" of sensation, but a musical perception. The sensation, however, which arises when the stimulus is a simple pendular-formed vibration is a convenient point of departure in the description of an element of sound, and we shall find in it the typical configuration of all tones however conditioned. At first glance this simple tone might be regarded as nothing but itself; yet when we observe it, and compare it with other tones somewhat differently conditioned, we find that our attention is caught by and focused upon several distinct aspects. These aspects are known to psychologists as the *attributes* of sensation. The analysis of a sensation begins, therefore, with a description of the attributes which integrate to form the sensation. For the ends of analysis we may, therefore, define a "pure" tone as a certain integration of all the attributes characteristic of the modality of sound, when each is used once and once only; the physical condition for this phenomenon being a simple pendular-formed vibration occurring within the audible range.[1]

PITCH

When we study the range of tones occasioned by simple pendular-formed vibrations, varying in frequency from 16 v.d. to 20,000 v.d., the first aspect to strike us is *pitch*. The slower frequencies occasion *low* tones; the faster ones *high* tones. If we consider the attribute of pitch alone, we find a gradual differentiation in height from one extreme of audibility to the other. Alterations of vibrational frequency that are just barely noticeable as differences of pitch are indicated by a fairly constant increment in the middle range of the scale. This increment may be given as approximately

[1] Cf. *81*.

one vibration per second; although under very favorable conditions the increment has been reduced as low as one-quarter of a vibration. But when the rate of vibration is less than 100 or more than 1,000 in the second, the increment must be rapidly increased if a noticeable difference of pitch is to be apprehended. The inadequacy of the ear for low and high pitch is most evident with rapid vibrational frequencies. At the rate of 3,000 v.d. the differential increment must be increased to about 10 vibrations; while at 4,000 v.d., and above, more than 40 vibrations may be required. These approximations are not to be regarded as definitely standardized results, for both practice and individual differences play a rôle in most of the tests which have been made on pitch, and it is very difficult to prevent other factors than pitch from entering consciousness to confuse the judgment. Since the object judged is a sound possessing many attributes, much practice in observation is necessary before one can inflexibly hold the attention upon a single aspect abstracted from the mass-effect of the whole tone. Thus, it may be questioned if the observers who record a *limen* less than one vibration per second have strictly adhered to the conditions of an attributive judgment. As in other fields of sensation, judgments of a sheer difference between two sensory impressions, though erratic, are often much finer than the limitation of a normal threshold for a particular attribute would seem to allow; the reason being that extraneous and variable moments in the total situation have afforded cues from which distinctions are drawn that have no bearing upon the precise difference which determines the threshold of a particular attribute. This shift of attitude on the part of the observer, from scientific precision to a common-sense judgment of difference, has brought many conflicting results with respect to liminal discrimination and the effects of practice thereon. As regards pitch-discrimination it is probable that no difference of height can be detected unless there is a difference of at least one vibration per second between the two generators of sound.

Despite these discrepancies, experiment has shown that for a considerable range of vibrational frequency the attribute of pitch, or height, varies with a fairly constant increment. In terms of vibration-numbers an arithmetical series is approximated when a series of tones appear to differ just noticeably in an orderly progression from low to high. During the course of this alteration of pitch the tone becomes neither *more* nor *less* in a quantitative way; it simply becomes different in its position within the qualitative (pitch-) series. One tone is "above" or "below" another, and we commonly refer to it as *higher* or *lower*.

This characteristic attaches to every sound, but it is more clearly apprehended in tones than in noises or vocables; indeed it has long passed as the most outstanding feature of a tone, though psychologists now consider the possibility of at least three other "qualitative" attributes, namely, *octave-quality*, *vocalic quality*, and *brightness*.

OCTAVE-QUALITY

Octave-quality derives from the observation that tones unlike in pitch may nevertheless be similar in kind. In a continuous transition from low pitch to high, the divergence of low and high constantly increases throughout the series; yet if we compare, not pitch alone, but tones of different regions in the scale, we remark that all tones whose vibrational frequencies are in the ratio of 1:2 bear a unique resemblance to each other. This is the familiar phenomenon of the octave. In some ways the passage of tones through the span of an octave is analogous to the color-series, since we pass from one tone through a transitional series to another that resembles it, just as we pass from red through the intermediate hues of the spectrum to purple, which is again reddish. But the analogy is incomplete, for whereas in passing from purple to red we regain our point of departure, and establish a complete circle, or closed series, the octave-tones are at once alike and different—they are alike as musical tones, but they differ in pitch. Furthermore, there are no obtrusive elements within the octave like the turning-points at red, yellow, green, and blue of the color-series. We therefore have, not a single continuum forming a closed series as in the case of colors, but the recurrence of likenesses and differences within a progression that is constantly altering its pitch. Every tone establishes an octave-relation with a certain tone above it and with another one below it in the series. Yet it never identifies itself with either, since the two tones of an octave always differ in pitch. For this reason we may for the present pass over the octave-quality as an elemental attribute of tone, our cursory examination having shown us that the octave always implies at least two tones in a perceptive configuration. Even if this configuration is as unitary an experience as any "pure" tone, the fact remains that the integration involves two pitches. We must, therefore, study the nature of the octave in connection with its musical effect before we can say whether or not it rests upon a unique quality inherent in a tone of a single pitch. Taken strictly by itself, a simple tone does not seem to warrant an inference regarding its octave, or regarding any other musical relationship.

VOCALIC QUALITY

We pass, then, to a second peculiarity of the tonal series, likewise described as qualitative. This is the resemblance which tones bear to vocalic sounds. The vowel sounds, according to familiar "Continental" usage—*u* as in "moot," *o* as in "mole," *a* as in "mart," *e* as in "mate," and *i* as in "meet"—readily find their places in an ascending pitch-series in the order in which they have been named. This has frequently been noted by investigators, and a considerable amount of experimentation has been undertaken for the purpose of determining the regions of pitch belonging to each of these different vowels.

A German psychologist, Wolfgang Köhler (*48*), had his

observers listen to a long series of pure tones ranging from low to high, while they were instructed to assign each vowel to its most favorable position in the series. His results were noteworthy because of the agreement independently reached by his observers; for the vibrational frequencies of the five principal vowels were found to occur with remarkable precision at intervals of an octave beginning with 264 v.d. for u, and ending with 4,224 v.d. for i. From these results Köhler concluded that a vowel-quality directly attaches itself to each of these five distinct regions of pitch, the intermediate or slurred vowels being assigned to positions between these regions. Köhler also placed the vocable m in the region of 132 v.d., while the sibilants s, f, and ch were assigned to three corresponding octaval points above i (49). Thus the regions of c in the scale come to assume a peculiar significance in Köhler's scheme, for, beginning with the octave below middle c which possesses an m-like character, we pass through middle c (c^1) which is u-like, c^2 which is o-like, c^3 which is a-like, c^4 which is e-like, c^5 which is i-like, c^6 which is s-like, c^7 which is f-like, and c^8 which is ch-like.

That the series of vocables involves an upward trend of pitch is not questioned, but Köhler's assignment of each vowel to a particular octave has been not only disputed, but virtually disproved.

In his preliminary tests Köhler employed a series of thirty tuning-forks ranging in vibrational frequency from 165 v.d. to 4,000 v.d. For an o-sound his three observers chose most

often the forks 435, 480, 550, and 600, which approximately
cover the interval of a fourth. For *a* they chose 1,100, 1,200,
and 1,365, an interval approximating a minor third; for *e*,
1,920, 2,000, and 2,400, also a minor third; and for *i*, 3,840
and 4,000, his two highest forks. An exact investigation of
these regions was then carried out with pure tones; the results indicating a selection of the octave-intervals as stated
in the preceding paragraph.

Whether this tendency to find an outstanding quality in
the c regions was determined by the instruction to observe
vocalic quality, or by a distinction, either natural or acquired, which inheres in these regions of the scale, is a question which can not be answered at present. Recent investigations of the vocalic sound, to be reported in the next chapter, show that the pitch of vowels is too indefinite to constitute an octave-series. It may nevertheless be true that the
c regions of the scale possess a qualitative nuance upon
which the selections of Köhler's observers were based (*123*).
As an attribute of the tone itself, these results point to an
outstanding quality of c-ness, rather than to an attribute
that can strictly be termed vocalic. We may therefore dismiss this quality from the tonal attributes, though we shall
return to the possibility of outstanding octaves when we
come to consider the phenomenon of "absolute pitch" (cf.
p. 162).

BRIGHTNESS

A third "qualitative" aspect of the tone is its *brightness*. When the pitch of a tone is clear and salient, we describe the tone as *bright;* otherwise it is dull. We must endeavor to set aside the more usual meanings of these terms, and regard brightness or dullness as simply a characteristic of the tone itself. Having done this, we find that in general all low tones are blunt or dull, while all high tones are sharp or bright. These aspects are sometimes described as the *mellowness* of low tones, and the *shrillness* of high ones. It is quite impossible to overcome the mellowness and bring out the pitch-salient of a low pure tone, and it is equally impossible to reduce the shrillness or saliency of pitch in a very high tone. This is especially noticeable in sounds which lie beyond the musical range. Because of the inertia of the sound-media when they are either very slow or very rapid in movement, neither a clearly ringing low tone nor a subdued and mellow high tone is ever heard. The uniform dullness of low tones is probably in part due to the slow-moving air, which is not sufficiently elastic to allow a single frequency to dominate the other frequencies of vibration that are set up along with it. With high frequencies, on the other hand, inertia being quickly overcome by the greater dynamic effect of rapid movement, the single frequency of the "pure" tone appears to shake itself free, as it were, so as to give a sharp effect to the sound as heard. This may account for the brightness or shrillness, the clear and ringing character, of all very high tones.

But is *brightness* an attribute?   There can be no question as to the appropriateness of describing some sounds as dull and others as bright.   Yet if these aspects are merely characteristics of the pitch-series, we have no need of a new attribute.   The experiments of Rich (*97*), which include a study of the discrimination of brightness, indicate that approximately the same differences of vibrational frequency will occasion a judgment of difference both in pitch and in brightness.   Rich, therefore, reached the conclusion that "pitch-brightness" is the proper term to employ for a single attribute the descriptive aspects of which, though different, nevertheless vary with the same increments of vibrational frequency.   Thus the inherent brightness of high tones and the inherent dullness of low tones belong together with the quality of pitch; whereas, if brightness were a distinct attribute, it must, of course, vary independently of pitch, and one should be able to detect differences of brightness between sounds whose pitch is the same.   Yet something of this sort actually appears to be the case when we compare tones with certain constituents of the vocalic sounds; for the most characteristic components of vocables are observably blunt and lacking in that peculiar saliency of pitch which is to be found in the tone.

Although we can not pursue this topic to advantage until we come to deal more explicitly with the perception of vocables (cf. p. 76 ff.), there is yet another line of investigation from which brightness emerges as an independent variable, with certain definite physical conditions governing its variation. In his study of the sound produced by very brief vibrational periods, Abraham (*3*) noted the clap-like noise which always accompanies them. With the aid of a siren, which we have previously described (p. 11), this investigator found that this clap-like noise accompanying brief tones possesses a distinct regional pitch, about an octave lower than the pitch properly belonging to the vibrational frequency. He also found that he could maintain a constant frequency, and yet secure tones of different character, by employing a disk rotating at a constant rate of speed while air was allowed to pass alternately through different series of holes of the same size and number arranged in concentric circles.

The sounds produced by different circles were phenomenally different; for though the pitch was constant, being occasioned by the same number of pulsations per unit of time, the tone produced at the periphery of the disk had a bright effect suggesting one of the higher regions of vocality, while the tone produced at the centre had a dull effect, and resembled the lower region of vocality. Since the difference in the sound-waves produced at the centre and at the periphery of the disk was not one of frequency, Abraham concluded that the phenomenal change in the sound must be occasioned by the *time* in which each single pulsation was operative. The moment of rarefaction involved in the passage of air through an opening is followed by a moment of compression which is somewhat shorter (cf. p. 11 f.) at the centre than it is at the periphery. Thus a modification of the wave-length is introduced without a corresponding modification in frequency. With this argument, Abraham concludes that vibrational frequency and wave-length are not universally interchangeable. In order to compute one from the other we must have a succession of waves unbroken by any pause. This is the case with most musical instruments, but it is not the case with the siren. Whenever the pulsations of air can be so modified that the wave's trough is somewhat longer in duration than its crest, the tone becomes *brighter*, although its pitch remains constant. Thus brightness takes its place among the attributes of sound as an independent variable. In any ordinary series of musical sounds brightness varies directly with pitch, because the conditions for the production of musical sounds are such that wave-length is usually a function of vibrational frequency. Rich's conclusion regarding the pitch-brightness of "pure" tones is therefore correct. But if the conditions of sound-production are such that the timerelation of rarefaction and condensation of the sound-pulse is unequal, this modified wave is accompanied by a degree of brightness that varies with the increased duration of one-half the wave-length.

The bearing of this discovery upon the interpretation of vocables and noises will be discussed in succeeding chapters. So far as the musical tone is concerned, *pitch-brightness* is the only "quality" we need accept. Let us turn then to a consideration of tonal intensity.

INTENSITY

In the usual classification of attributes, intensity, duration, and extensity are always listed. Evidently the first two of these are aspects of tone; for without reference to pitch or height we have loud and weak tones, and similarly, we have tones prolonged and tones of brief duration. While pitch is conditioned by vibrational frequency, intensity varies with the amplitude of vibration, and duration with persistence of stimulation and continuance of the neural response.

The analysis of intensity in sound depends upon a physical control difficult to secure; yet recent investigations by A. P. Weiss (*136*), seem to show that intensities can be compared when tones are of different pitch, as well as when they are of the same pitch; and Weiss has estimated that at least twenty-five differences of tonal intensity can be discriminated.

Another important feature of the intensity of sound is the inherent loudness of high tones, and the corresponding softness of low tones. This variation as it correlates with vibrational frequency is not, however, regularly graded from the low tones to the high. Max Wien (*138*), a physicist, who has investigated the normal acuity of hearing tones of different vibrational frequency, gives us a curve of sensitivity which shows that the inherent intensity of tones rises rather sharply from the lower range of audibility to sounds of about 1,200 v.d. Thereafter, sensitivity increases less rapidly, until at about 2,200 v.d. it begins to decrease, at first slowly and then quite rapidly, until the upper limit of audibility is reached. Wien's curve is plotted with reference to the energy required to produce audible tones in these different regions. With this objective basis, a tone at the maximal range of audibility (2,200 v.d.) appears to be about eight times as loud as a tone of 64 v.d. when both are produced by the same amount of physical energy. Although we are hardly justified in translating the intensity of stimulus into multiples of sensory intensity, the curve nevertheless is significant for the indications it affords as to the sensitivity of our ears to tones at different levels of pitch; and we may at least conclude that inherent intensity is the chief factor in determining the remarkable audibility of the sounds between 800 v.d. and 6,400 v.d., which are commonly employed in speech and in music.

DURATION

As for duration, the "going-on-ness" of a tone is a distinctly characteristic feature. It may be directly observed as an attribute of the pure tone, and it enters into many integrations which are important in the perception of music and speech. Thus duration is at the root of all meanings of *quantity* as this appertains to musical notation in "whole notes," "half-notes," "quarter-notes," etc. The duration of rests and the periods of time between tones, are likewise significant; for discrete temporal units of "filled" and "unfilled" time are ascribed to the perception of rhythm which means so much to us both in music and in language. As we shall see later, the lasting, or durative, character of noises, and, more especially, of vocalic sounds, is likewise a basis for the discrimination and perceptual interpretation of speech. Duration considered as an attribute of a simple phenomenal tone is, of course, quite another thing than the perception of time. What we observe directly is but the "going-on" of an experience which bears upon its face no reference to "before" or "after." Duration is transformed into time, when, in addition to the primary integration of duration with the other attributes of sensation, there is also a secondary integration of sensation with units of a higher and more complex order. In other words, "time" means something in practical experience which it does not mean for the analyzed aspect of consciousness that we call *duration*. Thus all objections to the acceptance of duration as an attribute which have been based upon the *relativity of time* are beside the mark. Curtis, who made an experimental investigation of this attribute, reaches the following conclusion (*22*, 45): "There are two ways of taking the temporal experience, as *progression* and as *length*. These stand at quite different levels, and are the results of quite different attitudes toward experience. A sensation, taken as it comes immediately to one, as it comes under a merely existential determination, progresses. The determination to compare or to estimate, however, tends to result in a taking of the experience as a length. Progression is the more ingrained, the more vital, aspect of the experience; without progression, length is impossible. Length is something that may or may not be added on afterward, and does not belong to sensation as such. That is, the sensation has length only in retrospect, has length only after it is over; it has progression while it is going on. In fact, the going-on *is* the progression, and by going-on we do not mean continuance in physical time, but the immediate experience of going-on. The length is, so to say, a 'fixing,' a 'making static,' of the progression for a simultaneous view. It is consciousness which involves supposing that the first of the tone is still there in some sense when the last of it comes. Now, to obtain this view, the progression must be referred to something outside itself, it must be given a definite beginning. As progressive alone, the experience seems to have no beginning or end. It is related no more to beginning and to end than the field of the closed eyes is related to definite points, say the farthest we can see to right and left; it is mere going-on, that is all there is to it. Length, on the contrary, is most clearly expressed as temporal distance between two points. It is the result of a perceptual, rather than of an attributive attitude."

It is in this sense of a unique "going-on-ness" that we use the term duration for an immediate attribute of sound attaching to all simple experiences in their bare existence as phenomena of consciousness.

EXTENSITY

There still remains to be considered the attribute of extensity, for all tones have an aspect which we can describe no better than by calling it a "spread-out-ness." Yet, just as in the case of duration, we must be careful to note that attributive *spread* is observed at a quite different level from that of a spatial judgment. The extensity of a color may seem to be more obviously related to its space-filling source than the extensity of a sound; because the perception of the color at once informs us of a shape and size to signify something like a flower or a wall, while the perception of a sound is not so trustworthy an index. By means of extensity in the field of vision we infer both the area and the depth of the object at which we gaze. In the case of tone we are unable to carry out a like perceptual interpretation. The small mass of a piccolo-tone is inadequate as a means of judging the size of the instrument, and large sounds are occasioned, not by the resonance of large surfaces, but by slow frequencies of vibration. It is, of course, true that by virtue of their greater inertia large bodies move more slowly than small ones, and therefore are likely to possess a slower frequency of vibration—which in turn affords a larger volumic effect. Similarly, a small body, that can vibrate more rapidly, will readily produce a tone of high pitch and small volume. But it is the same gross difference of frequency which occasions the change of pitch that likewise occasions change of mass, extensity, or volume; and although the two are intimately related as regards their physical origin, they nevertheless are psychologically distinct.

This aspect of massiveness or extensity of tone is commonly called *volume*. We say that some tones are voluminous, others less so. But precisely what does this mean? When carefully observed, the tone seems to revolve about its predominant pitch, and to spread therefrom in a vague surrounding nimbus of resonance. Perceptually we can neither define its limits nor determine its depth; for the edges of its pattern all seem to fade into nothingness. But we can, nevertheless, establish the perceptual contours of different tones with sufficient accuracy to compare the total mass-effects. And when we do so, we find, first, that judgments of volume are usually more crude than those of pitch, since a difference of pitch can be detected where there appears to be no difference of volume; and secondly, that discrimination of volumes seems to follow a geometrical progression of vibrational frequencies, whereas in the case of pitch the discrimination follows an arithmetical progression.

Thus the spread of a tone has no closer connection with space than has duration with time. Physically regarded, it is the vibrational frequency that appears to condition variations in volume, just as it also conditions, though in a different way, variations of pitch. There is nothing either bi- or tri-dimensional about the spread of a tone. It is a unique phenomenon of experience, and as such it has no geometrical function whatsoever. To what use its uniformities might

have been put in the perception of space, if this service had
not been so adequately taken over by sight and touch, we
have no means of knowing. As it stands, the perception of
tonal volume functions otherwise—as we shall see when we
have occasion to refer to it again in our study of music and
speech.

In his investigation of tonal attributes, Gilbert J. Rich
(97) has determined the *volume-limen,* or threshold of discrimination, with physically pure tones in three regions of
the scale, *viz.,* 275 v.d., 550 v.d., and 1,100 v.d. With the
aid of "interference-tubes" designed to eliminate all extraneous partials, an apparatus was provided capable of
giving physically pure tones. The tones thus produced were
very exactly controlled as regards their vibrational frequencies, and were conducted from a distant source through con-
duits to the ears of the observer in such a way that he could
hear successive pairs, varying in vibrational frequency by a
pre-arranged amount. The listener was then instructed to
observe the volume of two tones, and to determine whenever
a difference occurred. Series of experiments were performed
with five trained observers who agreed remarkably in their
estimations of volume-difference. In the lower region of
pitch, 275 v.d., about six vibrations were required to produce
a just noticeable difference in volume; in the middle region,
one octave higher, approximately twelve vibrations were
required; while in the upper region, an octave still higher,
twenty-four vibrations were indicated. Thus it was demonstrated that the discrimination of volume is determined

by a geometrical increase in the vibrational frequency of the sound, with a definite uniformity quite different from that of judging pitch; the limen for volume being about 1/46 of the vibrational frequency, while that for pitch in this region is a constant increment of approximately one vibration.

THE TONAL MANIFOLD

Even a simple tone, then, is a massive, or voluminous, entity, possessing a predominant central nucleus of pitch which seems to emerge as a sort of salient from the total mass of the sound. In addition, we have the loudness or intensity of the tone, and a persistence, or "going-on-ness," which is its durative aspect. With a gradual increase of vibrational frequency the pitch is raised and the volume decreased, though the heightening of pitch is more finely discriminated than is the attendant loss of volume. A graphic representation will perhaps aid in making these varied attributes clear (Fig. 9). A simple tone of momentary duration is here represented as a solid form, shaped like the turret of a mosque.

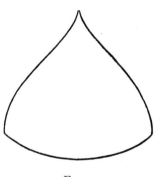

FIG. 9.

The circular base suggests the total volumic spread, though it is not intended to imply any special geometrical configuration. The point above is the salient pitch that dominates the sound, its sharpness being suggestive of its brightness, while, at the same time, the rise of the total mass from its base is intended to indicate the tone's inherent intensity. Were we to regard the whole mass as moving, the displacement would suggest its duration. The relative importance of pitch-salient and volume in determining the intensity of tones remains a subject for further inquiry.

We have already noted that high tones are inherently intense, and low tones inherently soft. It is also evident enough that all high tones are both sharp and lacking in volume, while the volume of low tones is often so great that the pitch can not emerge sufficiently to give saliency to the sound, which accordingly remains dull and regional.

To sum up: Pitch varies with vibrational frequency; the slower rate giving a low pitch, while as the rate increases the pitch grows higher. Amplitude of vibration occasions the intensity of the sound, and intensity may vary from more to less in every region of the pitch-series, though sounds in the higher range of pitch are inherently loud, while those of the lower range are inherently soft. Duration is occasioned by persistence of stimulation together with a continuing physiological effect upon the receiving organ and the auditory nerves. Volume, or extensity, follows the frequency of vibration, but in a different way from pitch; for whereas equal increments of approximately one vibration a second occasion a noticeable change of pitch at the middle range of the audible series, a proportional increase of .02 to .03 in vibrational frequency is requisite before we detect a difference

of volume between two tones. Low tones are most voluminous, while higher tones become gradually less so. As for
brightness, all low tones are dull, as they are also voluminous,
and all high tones are bright, as they are also small and
piercing, but in order to secure a variation of brightness

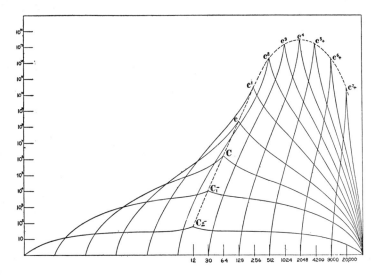

Fig. 10. The ordinates of this diagram are to be read as a logarithmic
expression of auditory sensitivity. Wien's measurements have shown
that in order to be audible a tone of 50 v.d. must possess 100 million
times the energy of a tone of 2,000 v.d.

that is independent of pitch, we must have conditions
whereby the wave-length is so modified that its to-and-fro
half-periods are unequal without a corresponding alteration
of the vibrational frequency of the complete wave.

In Figure 10 an attempt is made to represent the tonal
manifold graphically, by indicating each of the more important features of a series of pure tones progressing through-

out the range of audibility (cf. *82*). The figure must be accepted with a certain reserve, because the data from which it is drawn, while largely factual, are not entirely so. The following description will indicate such assumptions as have been made. Each tone is pictured as in the previous figure, with a certain spread on the base-line to suggest its volume, from which it rises to a peak suggesting its pitch-brightness. The height of each peak conforms to Wien's curve of sensitivity, thus indicating the inherent intensity of tones at every level of the scale. Duration, since it involves movement, is not included in the scheme.

It will be noted that the total spread, or volume, of the lowest audible tone comprises within its range the volumic emplacement of all the higher tones, the extreme upper point of emplacement being identical for every tone. This assumption is in accord with Watt's theory, and has been tentatively accepted on the grounds he has advanced (*134, 62* f.).

Decrease in the spread or size of volume by a fixed amount for each octave is assumed throughout the musical range of 64 v.d. to 2,048 v.d. This is done so as to indicate the apparent constancy of the octave-intervals. The theoretical justification for this assumption is developed in a later chapter. Both above and below this range, however, the fractional increase of 1:2 is presumed to vary. The volumes of the lowest tones being relatively greater than the normal increase would warrant, a larger fraction is needed to measure an interval of equal volumic proportion. This accords with the observation that low tones appear higher in pitch than they should do with reference to their vibrational frequencies (*59*). Similarly the highest tones decrease in volume more rapidly than tones in the middle range, and hence again a greater proportional variation than 1:2 is indicated for the octave. This also agrees with the observation that tones of the four-accented octave, and above, sound flat. The total range of volume has been divided, somewhat arbitrarily, into 228 steps, each step representing a discernible interval as determined by a clearly-defined difference of volume. These intervals as they occur within the range of musical tones approximate a quarter-tone step. The volumic spread of each octave is shown as diminishing at the constant rate of 24 quarter-tone intervals, which approximates the threshold of Rich's volumic discrimination. In the highest and lowest ranges of the scale the judgment of intervals is so uncertain that the two lower and the three upper octaves are but sheer guess work. All we know of these regions is that tones below 40 v.d. appear to be a little higher than would be warranted by the rate of vibration, while in the upper range, tones of 3,000 v.d. and above are flat, and at about 4,000 v.d., according to von Maltzew (*59*), the accurate judgment of intervals breaks down completely. Thus equal decrease of volume as dependent upon equal vibrational ratios has been assumed with octaves and other musical intervals only for tones within an approximate range of 50 v.d. to 3,200 v.d.

Turning now to the pitch of tones, this is indicated by

the central point, or salient, in the upward-rising mass of volume. It will at once be noted that as the volume decreases the pitch becomes sharper, which suggests the aspect of brightness. As pitch rises, it emerges more and more clearly—becomes more and more salient. The upward trend from the base-line likewise suggests the variation of inherent intensity attaching to tones at different pitch-levels, and the curve which describes the salient peaks of these progressive tones is the one already referred to as having been determined by Max Wien in his study of auditory sensitivity for tones of different pitch.

A special interest attaches to this curve of sensitivity, because of the indication it gives regarding the threshold of pitch at different levels of the scale. In the lower range, successive tones coincide to so large an extent that the sensitivity to pitch may be only slightly greater than the audibility of volumic differences. The pitch-salients of low tones are barely distinguishable in the graph; an appreciable distance, or interval, being requisite before one pitch emerges distinctly from another. In the higher ranges this is not the case, for with salient tones one pitch distinguishes itself from another even though there is no perceptible volumic difference upon which a judgment of interval can be based. Thus the number of discriminable pitches within an octave has been shown to increase steadily until we reach tones in the region of 2,000 v.d. when it begins to decrease. Decrease of sensitivity in the upper range, together with an increasing inability to judge volume accurately, both correlate with the difficulty of discriminating pitch in this region. Since, however, the absolute difference of vibrational frequencies required for a given interval is progressively greater as the vibration becomes more rapid, there still may be a larger number of discriminable pitches in the interval at the higher levels than we find in a like interval at a lower level of the scale. Thus the distance between the tones of an interval at a high level of pitch may seem greater than the distance between the tones of the same interval at a lower level, because a greater number of differentiable pitches are traversed (cf. Stumpf *122*, I, 127) in the former case than in the latter. Our figure represents, then, the progression of tones throughout the range of audibility, and tries in a general way to indicate the course taken by volume, pitch-brightness, and intensity throughout the tonal manifold.

CHAPTER IV

VOCABLES

SCIENTIFIC INTEREST IN VOCAL SOUNDS

LET us now turn to the analysis of our second class of auditory perceptions, the sounds of vocal utterance. Since a later chapter will be devoted to language, our present interest attaches merely to the sounds of the human voice as a separate class of sense-perceptions.

Vocal sounds can be roughly divided into vowels and consonants. As already noted, vowels possess characteristics which closely ally them to tones, whereas consonants are more like noises. The vocables have been investigated in numerous ways and from various sides. They interest the linguist as elements in the evolution of language, and the teacher in planning an adequate course of instruction in a foreign tongue. Linguistic, etymological, and phonetic research have all contributed to our knowledge of the sounds that occur in speech. What we know about linguistic sound is therefore largely a result of investigations undertaken to clarify the history of language, and to comprehend the possibilities and the limitations of vocal utterance.

Not only the linguist, but the physicist also has found a special problem of acoustics in the sounds produced by the larynx and the buccal cavity. His mode of analysis naturally has been that of measuring the sound-waves of speech. Exact investigations of the complicated forms of vocal resonance, with due regard both for its frequency and its amplitude, have therefore been undertaken.

The physiologist and the otologist are likewise concerned in the function of vocal production and its apprehension; for the action of the vocal cords, the movements of the tongue in speech, and the formation of the resonating cavities of the mouth and nasal passages, are all physiological problems. In the treatment of diseases of the ear, deafness to speech assumes a special importance in the practice of the otologist. Accordingly, the nature of the vocables interests both the student of physiology and the medical practitioner; the latter being concerned not alone with defects of hearing, but also with defects of articulation, faulty speech, imperfect phonation, etc.

Finally, the psychologist finds in the vocables a variety of auditory perception possessing certain unique features of experience. It is to these features that we shall here direct our attention; and first of all to the sound of vowels. This subject has already been touched upon, both with reference to Köhler's attribution of a "vocal quality" to tones, and also with regard to "brightness" as an independent attribute of sound.

ATTRIBUTIVE ASPECTS

Every sound, the vocables included, possesses each of the four attributes of pitch, intensity, duration, and extensity;

and the description already given for tone will apply equally to all types of sound.

This statement is perhaps most obvious as regards intensity and duration. We can conceive no sound, however complicated its structure, however varied its meaning, that will not evince an aspect of loudness or softness on the one hand, and an aspect of brevity or continuance on the other. In comparison with intensity and duration the meaning commonly assigned to pitch is more equivocal; for we ordinarily do not speak of the pitch of a noise, nor always give serious attention to the pitch of a voice; yet if we define pitch as height, this aspect is never absent from any sound whatsoever, though it may easily be obscured, as, for instance, in a complex sound which embraces a large number of different pitches. But even in the case of a sound produced by a simple pendular-formed vibration of slow frequency, the pitch is never salient; bluntness or dullness being a universal characteristic of all low-pitched tones. Yet indefinite though it may be, pitch is always an observable feature of every sound.

Volume is also attributive to every kind of sound, whether it be simple or complex, and, like pitch, it too depends upon the frequency of vibration. But whereas different pitches may be heard simultaneously in a musical chord, the volume of the chord seems always to be that of its lowest and most voluminous component. Thus the volume of any sound is in large measure, if not entirely, determined by the volume of the lowest component of the total sound-mass.

Since the perception of a vocable is that of a distinct configuration of sounds, each possessing all the four above-mentioned attributes, the questions arise: What is the pattern of
this configuration whereby the vowel is distinguished from
the perception of a tone, and are any other attributes than
these four concerned in distinguishing the vocalic pattern?

As already pointed out, a comparison of vowels and tones
emphasizes both their likeness and their difference as regards pitch. The vowels readily assume a definite order in
a rising scale of pitch, and characteristic regions of pitch
have been assigned to each of the vocalic sounds. Yet a
tone in one of these regions and its corresponding vowel are
by no means identical. What, then, is the difference between
them? Answers to this question have been given by a number of investigators, but only quite recently have the
analyses been carried far enough so that we may venture an
opinion as to the probable difference between these two types
of sound.

THE SCIENTIFIC STUDY OF THE VOWELS

As early as 1837 Wheatstone, the English physicist, advanced a theory of vowel-tones which he based upon some
contemporary experiments made by Willis. Later Helmholtz attacked the problem, and reached an analogous conclusion, which was that "the vowels of speech are in reality
tones produced by membranous tongues (the vocal cords),
with a resonance-chamber (the mouth) capable of altering

in length, width, and pitch of resonance, and hence capable
also of reinforcing at different times different partials of the
compound tone to which it is applied" (*35*, 103). According to Helmholtz's observations the pitch of the five chief
vowels can be located at the following points of the scale:

> *u* at f,   175 v.d.
> *o* at bb¹,   466 v.d.
> *a* at bb²,   932 v.d.
> *e* at b³,   1,976 v.d., and in addition a lower resonance
>   at f¹,   349 v.d.
> *i* at d⁴,   2,349 v.d., likewise with a lower resonance
>   at f,   175 v.d.

This is sometimes known as the partial-tone theory, since
the vowel-sound is found to depend upon the presence in the
voice-tone of one or more partials which resonate strongly
in a characteristic region of pitch.

Another investigator, the physiologist Ludwig Hermann
(*36, 37*), takes exception to Helmholtz's assumption that
the vowel-tones must be harmonics of the fundamental tone
upon which they are uttered. According to Hermann the
vowels occur in fixed regions of the scale, and constitute a
peculiar structure which he calls a *formant*. The formant
is not a phenomenon of resonance, but is produced in the
mouth-cavity by an act of blowing, and is independent of
the tone that issues from the larynx. Thus, according
to Hermann, any vowel may accompany any tone which
the voice sounds. It is, as it were, a semi-independent utter-

ance, attributable to the resonating capacity of the mouth in appropriating a portion of the energy that is being emitted as a tone from the throat. Whereas by Helmholtz's explanation the vowel can be produced only when a partial tone of the fundamental chances to fall within a characteristic region of resonance, Hermann believes this condition to be unnecessary.

Neither Helmholtz nor Hermann makes use in his measurements of the refined technique employed in the later work of Dayton C. Miller (70) on the physical side, and Carl Stumpf (118) on the psychological. From the investigations of these two, supplemented by the important, though less fundamental, work of E. R. Jaensch (42) and of Heinrich Schole (102), we are now able to reconcile the chief differences in the views held by earlier investigators.

In brief, the most recent results on this subject point to the following conclusions: Vocalic sounds all fall within the harmonic scheme of resonance, just as Helmholtz had suggested; but the characteristic phenomenon of the vocable is a unique structure, to which the term *formant,* used by Hermann, is quite appropriate. The fundamental tone on which the vowel is sung is not a negligible factor, and if it chances to be too high in pitch to include the particular region of resonance needed by a special vowel, that vowel fails of utterance. Often, however, the fundamental is entirely lacking as a physical vibration, being supplied to the sound-perception only as a difference-tone of the characteristic partials. On the other hand, the formant is something more than a mere partial tone in a certain region of the scale. Though an exact phenomenological analysis of this interesting sound-structure has yet to be made, the evidence already at hand points toward a physical counterpart in the distribution of energy over a certain group of receptors. In their combined response these receptors give rise to the peculiar smoothness or dullness attaching to the formants of a vocal utterance. And hence it may perhaps be inferred that not one but a number of adjacent receptors of the ear are being aroused.

Although complete agreement does not exist regarding the exact position of the formants in the scale, the discrepancies have been no greater than one might expect in view of the fact that vowel-sounds are transitional, and that innumerable intermediate vocables fill the gaps between each two of the "pure," or typical, vowel-sounds. Indeed, the agreement among the more recent results of Stumpf, Miller, and Schole is very striking; and particularly in the comparison of formants as determined by Miller and Stumpf. In this connection it should be noted that Miller was working with vowels of the American idiom, and that Stumpf was using the German phonation.

While their studies were entirely independent, and their methods quite different, each of these investigators assigns a double formant to u (oo) with the region of greatest resonance lying below g^1 (400 v.d.). According to Stumpf this formant is usually given with the fundamental tone upon which the sound is uttered, and it may range as low as c

(128 v.d.). In addition to this fundamental, both investigators find that u may also contain a characteristic formant in the region of g^2 (800 v.d.).

For o Stumpf finds a single formant at g^1, embracing one or more overtones. Hence the same pitch appears to act differently when it appears as the fundamental of the voicetone in u and when it appears as an overtone in o. Miller's o-formant is slightly higher than Stumpf's, culminating at a^1 (461 v.d.). The vowel a embraces a number of more or less distinct vocalizations, varying from the broad a in aw and ah to the more acute a in at, and even approaching et. The first of these has a primary formant whose pitch lies between f^2 and c^3 (700-1,000 v.d.), with an additional formant sometimes appearing at d^4 (2,400 v.d.). According to Miller and Schole the higher a's shift their formants as high as e^3 (1,300 v.d.); and Miller finds a double formant for at and $et;$ the first at g^2 (820 v.d.) and bb^3 (1,840 v.d.); the second at f^2 (700 v.d.) and b^3 (2,000 v.d.), respectively.

The double formant is also characteristic of the vowels e (mate) and i (meet). According to Stumpf the primary formant of e falls together with the upper formant of a, at d^4; the secondary formant is at g^1, as in o. Miller's results are the same, except that he finds the secondary formant ranging somewhat higher in the scale. For i Stumpf finds formants at e^4 (2,600 v.d.) and a^4 (3,600 v.d.), associated with a fundamental ranging from g^1 downwards as in the case of u. Miller's formants cover the same regions of resonance, though he finds but one upper formant centring about

e⁴ (2,640 v.d.). The last-named difference is illustrated by
the accompanying photographs which Professor Miller has
supplied (Fig. 11). They are the records of *u* in "room"
and *i* in "bee," with approximately the same fundamental
tone of 165 v.d. A physical analysis of these two sounds
shows that their principal difference is a partial tone in the
neighborhood of 2,500 v.d., which is present in the "ee"
sound and absent in the "oo" sound.

In support of the conclusions reached, some reference
should be made to the experimental technique employed by
these investigators in securing their results. Miller, who
approaches the problem as physicist, has invented a very
ingenious apparatus, the *phonodeik,* which may be employed
in visualizing sound-waves. One can stand before this
instrument and utter a vowel-sound which will immediately be projected in the instrument as an illuminated
curve upon a dark background as shown in Fig. 11. By
varying the pitch of the voice until it attains the proper
region of the vowel uttered, the curve, at first complicated,
resolves itself into a simple sine-wave.

These projections can be photographed, and with the aid
of a mechanical analyzer even very complex curves can be
resolved, in accordance with the principle of the Fourier
analysis, into their partial components. When this is done,
the amplitudes of the various components indicate which
were the characteristic regions of resonance in a particular
vocal utterance.

Stumpf's analysis was made with the aid of interference-

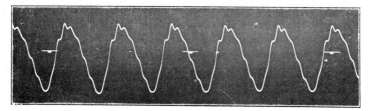

The vowel "oo" in *room;* fundamental frequency 169 v.d.

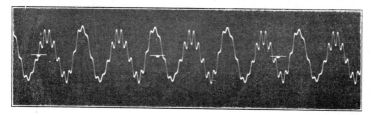

The vowel "ee" in *bee;* fundamental frequency 165 v.d.

Fig. 11.

tubes carefully adjusted with regard to all the partials of the fundamental tone upon which the vowel was being spoken or sung. His apparatus permitted him to include or exclude at will the various partials of the voice-tone. He was thus able to determine just which partials were characteristic of the vowel, and which ones were accessory or irrelevant. After determining the formants in this manner, Stumpf performed a painstaking synthetic experiment in which pure tones were assembled in the number and intensity necessary for the production of various vowels. The outcome of these syntheses was successful: a comparison of the synthetic vowels with those of the human voice showed that the artificial vowels were even more satisfactory to a large number of observers than were the corresponding productions of a natural voice heard under like conditions. In this way Stumpf appears to have effectually disposed of Hermann's contention that the vowel-sounds are inharmonic with reference to the fundamental tone of the voice; for neither in his analytic nor in his synthetic experiments did Stumpf find any indication of non-harmonic components.

THE NATURE OF THE FORMANT

But although vowels have been thus successfully analyzed into the partial tones of a complex clang, the nature of the formant which characterizes each vowel is still somewhat obscure. Both Miller and Stumpf have shown that each formant occurs within a fixed region of the scale. The average extent of this region is estimated by Stumpf as approximately the interval of a minor third, while Miller's measurements allow it a somewhat greater range. We have yet to consider whether the characteristic sound of the formant owes its vocality merely to the pitch of this region, as Köhler maintained, or to a resonance of the mouth-cavity which increases the amplitude of all partials that chance to fall within the region of the formant (*1*). Stumpf has shown, however, that a partial tone is not alone sufficient to produce a vowel; for a certain number of partials is always involved, and each successive vowel in the rising series of pitch has its foundation in some other vowel, as *o* in *u*, *a* in *o*, *e* in *o*, and *i* in *u*. Since at least two partials are always present in any vowel-sound, the phenomenal structure of a vowel must be more complex than that of a "pure" tone. The difference between a single formant and a simple partial tone occurring in the same region of pitch is the problem now before us.

An ingenious series of experiments conducted by E. R. Jaensch (*42*) appears to demonstrate that a pure tone produced by a single vibrational frequency grows increasingly vocalic in quality whenever, instead of a single sine-wave from the generator, a mixture of sine-waves was produced by adding together different frequencies of vibration whose mean variation was slight. Thus, in his synthetic experiments, Jaensch reports the production of sounds of vocal quality by the simple expedient of adding to a single vibrational frequency a small number of but slightly varying vibrational frequencies. As the mean variation increased beyond a certain limit, the sound became noisy. The source of Jaensch's sound was a beam of light playing upon a *selenium cell* electrically connected with a telephone-receiver. The light was interrupted by a revolving disk with a hatched edge which produced the desired frequencies. By a combination of disks the experimenter was able to secure the mixed sine-waves which he found to be requisite in the production of vowel-sounds.

These experiments have recently been repeated by Hans Lachmund (*54*), one of Jaensch's assistants, who finds that not only variation of wave-length, but likewise variation of amplitude, together with the periodic fluctuations occasioned by both these variations, is calculated to modify a tone in the direction of vocality. In particular, the periods of recurrence introduced by the modifications of pitch and amplitude introduce a supplemental low tone which serves as a fundamental. It was found that variations of amplitude alone were as effective as variations of wave-length, so that the essential condition for the vocalic sound would appear to be a disturbance of the "pure" tone. Not all his observers could readily apprehend the vowel-sound, but when they did so, abstraction from the pitch of the original tone was necessary. The disturbance of pitch seemed to resolve the sound into a vowel by causing the tone attributable to the frequency of the objective vibration to disappear. The "voice-tone" aroused by the periodicity of interference then became the vehicle of the vowel-sound, replacing the true

pitch. These experiments demonstrate that the regional frequencies of the formants of *o* and *a* are constant, and approximately identical with those assigned to these vowels by
Stumpf. But it was found possible to vary the voice-tone
from 90 v.d. to 100 v.d. for *o*, and from 128 v.d. to 305 v.d.
for *a*, a result which appears to be at variance with Stumpf's
conclusion that the resonating region must be a harmonic
of the voice-tone, and in agreement with Hermann's contrary views.

The question has been raised whether a telephone-disk is
capable of resonating in such a manner that the complex
electrical impulses received under the conditions of Jaensch's
experiments could all be perfectly registered in the sound
heard. Lachmund informs us that care was taken to see
that the disk vibrated in sine-waves, but the information
whether it was capable of reporting to the ear each component in such a mixture of sine-waves is not vouchsafed.
In view of the fact that the voice-tone, a new sound produced by the period of disturbance, was found to be an
essential feature in the synthetic vowel, we may perhaps
conclude that the vocable arises not only as a result of a
disturbance of pitch, but also by the addition of a fundamental tone with which this phenomenon of disturbance is integrated. Lachmund admits his uncertainty as to how much
of the vocalic effect is attributable to the factor of disturbance of the original sine-wave, and how much to the presence
of a voice-tone. With the results of Stumpf and Miller before us, we may be reasonably sure that the vocable is a

complex integration of sounds, in which a fundamental possessing tonal quality is always given. But, in addition, one or more formants must also be present, and according to Jaensch and Lachmund the condition under which a tone is changed into a formant is a disturbance of what would otherwise be a simple sine-wave. This disturbance was experimentally conditioned either by the accretion of sine-waves of slightly varying frequency, or, with a constant frequency, by a slight variation of amplitude in the succeeding vibrations. The production of the voice-tone through the periodic recurrence of the disturbance of the sine-waves was overlooked by Jaensch in his original experiments, and hence he attached undue weight to the mere production of mixed sine-waves. This error in judgment is corrected by Lachmund.

In the synthetic experiments of Stumpf and Miller no such disturbance in the regions of the formants was introduced. Shall we conclude, then, that the disturbing factor of the Jaensch experiments was only the fortuitous means of creating a voice-tone to which the modified sound of the original frequency could be referred as an overtone?

Since a precise analysis of the formant has yet to be made, it would be unwise to go too far in generalizing from the data at hand. But assurance seems to be given that the vocables are perceptually complex, and involve both a voice-tone or fundamental, and also a series of overtones among which there must be included sounds emanating from certain definite regions of the scale. These regional sounds constitute the formants of the different vowels and consonants. When measured by the average frequency of their range, the regions that mark these formants are not necessarily harmonics of the fundamental voice-tone. Yet if we consider that the range is a minor third or more, it can well embrace a partial which is harmonic to the fundamental. When Lachmund insists that the formant need not be a harmonic of the fundamental, he is measuring his formant by the frequency of the sound which has been disturbed, with which the frequency of the voice-tone, since it is dependent upon the period of the disturbance, has no necessary harmonic relation. Thus the independence of the voice-tone and the original frequency-tone is evident. But if we assume that the disturbance has been productive of a regional resonance which replaces the original frequency of the simple sine-wave, this region might be extensive enough to embrace partial tones that do harmonize with the voice-tone produced by the period of the disturbance. We may therefore question Lachmund's inference that because the vowel remains the same while the voice-tone is being varied, it follows that the formant and voice-tone must be inharmonic. Since one of the conditions under which the vowel was perceived was a shift of attention from the original sound occasioned by the frequency-number to the voice-tone introduced by the periodic disturbance, the only warranted inference is that a regional resonance lacking a definitely observable pitch has replaced the tone with which the experiment was begun. Indeed, those observers who retained their original attitude continued to hear a tone, and not a vowel, even after the disturbance had taken place. Baley (6), when experimenting upon the simultaneous perception of a number of pure tones that varied but slightly from one another in pitch, reported that his observers identified the pitch of such a combination as being that of its *mean tone,* but he has nothing to say of a vocalic effect. Only when Lachmund's observers were able to abstract from the frequency-tone by shifting their attention to the newly created fundamental, or voice-tone, did the vocalic effect appear.

Both Stumpf and Miller have shown that every region of the scale from c to c [5] resonates noticeably with some vowel or other; and every typical vowel has one or two characteristic regions of resonance within this range, without which the vowel in question can not be sounded.

Lachmund records numerical data for but two vowels, *o* and *a.* As to the former, he states that the voice-tone may vary from 90 v.d. to 150 v.d. while the vowel remains at the frequency-number 450 v.d. to 465 v.d. But 450 v.d. is the fifth partial of 90 v.d., and the third of 150 v.d. In the case of *a,* frequency-numbers of 915, 930, 945, and 960 v.d. were found to correspond with period-numbers of 305, 248, 189, and 128 v.d., respectively. As we have already noted, there are several different kinds of *a*-sound with a primary formant that varies at least from 700 v.d. to 1,000 v.d. Each of the above-named period-numbers would find a partial tone within this region. As a matter of fact, the frequency-numbers in the first and third instances named are exact overtones of the fundamentals assigned to them, while in the other two the discrepancy noted is only approximately that of a semitone.

We are therefore unprepared to conclude that under natural conditions of vocalization the vibration taking place in the region of the formant can ever be inharmonic to the fundamental. Certainly Lachmund's results offer no legitimate support to Hermann's view that the formant is unrelated to the voice-tone; for it must be remembered that the voice-tone of the Jaensch experiment is a fortuitous adjunct to an artificial formant created by the disturbance of a sine-wave. A vocalic effect may therefore have been produced in which the normal relationship of a fundamental to its partials was only approximated. Hermann's theoretical interest turned upon the mechanism of the throat- and mouth-cavity in the production of vocables, while that of Stumpf and Jaensch is primarily the analysis of the vocal sound. As we have here shown, there is no serious disagreement in the results of Stumpf and Jaensch, although they were secured by quite different methods.

A further point mentioned by Stumpf seems to support Jaensch's association of the formant with a disturbed or regional effect. Stumpf found that when but one partial constitutes a formant, the vowel-quality is less marked than when there are several partials falling within the specified region. Vowels, therefore, are more effective when they are uttered at a low pitch which can supply a number of partials to constitute the formant; from which we infer that the formant is normally conditioned by a regional resonance, rather than by a simple partial tone of extraordinary amplitude. Although Stumpf leaves open the question as to how the physical intensity of the partial tones which constitute the formants is related to the physiological response of the organ of hearing, Miller has expressed a private opinion that the occasion for the modified sound we call a vowel is a distribution of kinetic energy over a certain region of the basilar membrane, rather than a summation of amplitudes of the separate partial tones which are sounded.

Some evidence as to the structure of the formant is given by the recent experiments of Abraham (3) on certain types of noise. The *Knall*, or clap-like sound, to be observed in a siren-tone of very brief duration was noted to possess a resemblance to a vowel which varied with the wave-length, as measured by the absolute time of pulsation, but not with the frequency. Thus, with a constant speed of the motor which gave a tone of g on different rings of equally sized but unequally spaced holes of the siren, so placed that the same number of holes occurred in each ring, the sound of the outermost or largest ring was like *a* in *mate* while the sound produced by the smaller circles varied with their size through *a, ao, oa,* and *o,* to *u,* which last was produced with the smallest circle. This modification of the tone is described as one of *brightness* or *vocality*. Of the two, brightness seems to us the more appropriate term, for as Abraham also remarked—agreeing with Lachmund—a difference in brightness was very easily apprehended, while the determination of the vowel was much more difficult, requiring a direction of attention upon vocality, and away from pitch.

The tones of the siren are not pure, but rich in partial vibrations, and Abraham has found that the "primary" wave-length is responsible for variations of brightness and vocality, with reference not only to the fundamental, but also to its partials. In his experiments with interference-tubes calculated to eliminate all tonal components, both of the fundamental frequency and its partials, the primary wave could still be heard, together with certain partial wave-lengths, with intensities that did not correspond with the Fourier reckoning.

A THEORY OF THE FORMANT

May we not conclude, then, that the formant, which is so characteristic of the vocable, is a unique sound-structure, limited in its production to a certain region of resonance appropriate to the particular vocable, and phenomenally distinguishable from the ordinary partial tone by the absence of a salient pitch? In the experiments of Jaensch and Lachmund a tone is modified by a disturbance which effectually destroys its saliency; the frequency-tone therefore falls into the background of consciousness, and becomes a dull accompaniment of the voice-tone which emerges to take its place as the fundamental tone of a perceptual integration. In the synthetic experiments of Stumpf and Miller a similar effect was produced, not by the obvious means of disturbing the frequency, or varying the amplitude, of successive waves, but by increasing the resonating energy of certain partials. As a result, their relation to other partials of the total clang was no longer that of an ordinary musical sound in which the partials each have the effect of a simple sine-wave upon their receptors. Instead, the energy attaching to the critical sounds of the formant regions was increased to such an extent that several receptors adjacent in the scale were aroused to produce a dull or softened sound, lacking the salient pitch that characterizes the partials of an ordinary musical clang. In both cases the reduction of brightness necessary to change the sound from a tone to a formant may perhaps be attributed to a disturbance of wave-length. If our inferences are correct, the sounds of the human voice are but modified clangs, the modification being caused by the peculiar mechanism of voice-production and the resonance of the mouth and throat. Resembling in part, as it does, the production of siren-tones, the voice gives ample opportunity for modifying certain wave-lengths so as to occasion variations of brightness that are independent of pitch. As compared with other musical instruments, the voice pro-

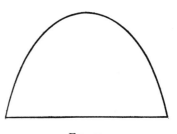

Fig. 12.

duces these unique sound-structures, the formants, which can be described as dull tones occurring at a place in the scale where tones are not ordinarily dull but bright. A graphic representation is here given (Fig. 12) to suggest the structure of the formant as compared with the structure of a pure tone represented in Fig. 9.

In other words, when a tone suffers a disturbance of its evenly recurrent pulsations, it loses the characteristic brightness that belongs to its salient pitch, and becomes a formant. If the disturbance is slight, the pitch-region is still discernible, although the sound has become smoother and duller. If the disturbance is increased, the vocality of the formant is lowered, until finally it disappears in a noise. The maintenance of the formant in a definite region of pitch to produce a certain vocable is enhanced by the attendant voice-tone, or fundamental, which by its partials serves to reinforce the region in which the disturbance is taking place.

These are, of course, but inferences which seem to be warranted by the data at hand. Before the interpretation is complete, or indeed assured, we must, however, have a deeper insight into the phenomenology of the formant and into the conditions under which it appears. A satisfactory description of the formant may be difficult to gain, for apparently it can not be observed in the focus of attention. However, the independent variability of brightness as a function of wave-length furnishes a very important datum by which we can now understand how a difference is possible between the essential components of vocables and tones. We have therefore a systematic setting for a unit of sound comparable with the "pure" tone, the chief variation therefrom being indicated by a decreased brightness which correlates with a lengthening of the crest as compared with the trough of the sound-wave.

Brightness would then be independently variable with respect to the distribution of the kinetic energy involved in the production of sound. When the energy is concentrated, as it were, upon a single receptor—be this in fact a single resonator of the basilar membrane, or a small area of the swinging tissues of the inner ear—we would have the conditions requisite for the production of tones that are intrinsically "bright." But when the energy is distributed so as to engage a number of adjacent receptors, or a wider range of basilar tissues, the resultant sound, while retaining the same pitch, would have become duller, mellower, less tonal, and more vocalic.

By rendering the pitch dull instead of bright we have introduced a new pattern of integration—the formant—which, when perceived in an appropriate complex of fundamental tone and overtones, gives us the configuration of the vocable. We have no need for an attribute of vocality, which would indeed be a misnomer, since all vocables are perceptual integrations; but the independent variability of brightness affords a means of characterizing the formant in attributive terms, the formant having already been shown to be a necessary constituent of the vowel-sound.

But we must not permit an over-hasty generalization to outrun the facts of the case. We have taken the formant to be a primary integration of all the attributes of sound, though at present we know little more of its phenomenology than that its brightness is less than the brightness of a "pure" tone of the same pitch. The variations of volume, duration, and intensity which may likewise enter into this integration have not yet been worked out.

CONSONANTS

In turning from vowels to the other type of vocable, the consonant, we enter a field of perceived sound which on account of its greater complexity is less well-defined. Yet the studies of Hermann (*36*) and Stumpf (*126*) furnish evidence that the structural difference is chiefly one of complexity. Consonants are more noisy than vowels, and their peculiar patterns are therefore less dependent upon the harmonic series of partial tones which constitutes the clang of any vocal utterance. Yet apart from their noisy constituents, which are the subject of special consideration in the next chapter, consonants likewise include formants, and we need assume for them no attributes other than the ones already set forth.

According to Stumpf the half-vowels and consonants can each and all be analyzed in the same manner in which he has so successfully analyzed the vowel-sounds. From the experiments he reports, the formants of the sibilants rise in pitch in this order, *sch, s, f,* and *ch* (palatal), as Köhler had previously demonstrated. Yet Stumpf does not find that these sounds require the very high frequencies Köhler

thought necessary, nor does Stumpf, like Köhler, place the
m-formant below the *u*.

VOICED AND UNVOICED VOCABLES

In considering the consonants in their relation to the vowels, a distinction should first be drawn between voiced and
unvoiced vocables. We have seen that the vowels have
much that is tone-like about them, and this tonal character
is of course more perceptible when the sound is "voiced,"
than when the same vocable is whispered. We can secure
a more accurate determination of the formants, both of
vowels and consonants, if we study the unvoiced or whispered sounds as Stumpf has done in his most recent contributions to this subject (*125, 126*).

In general, the range of the formants for unvoiced vocables
was found by Stumpf and Miller (*126; 70,* 235 f.) to be
somewhat higher for the vowels having a single formant
(*u, o, a*), and lower for those having a double formant
(*ö, ä, ü, e, i*), than it is for the same vocables when they are
voiced. Apparently the lower tonal components of the voice-
lang have the effect of reducing the apparent brightness
of the lower formant-regions, and of increasing that of the
higher regions. Yet the shifting of the formant is not so
great as to suggest a structural difference, and since whispered vowels are simpler sound-structures than voiced
vowels, we can perhaps place more reliance upon the posi-

tion of the formant thus indicated than we can when the sound is accompanied by the complexities of a complete vocalization.

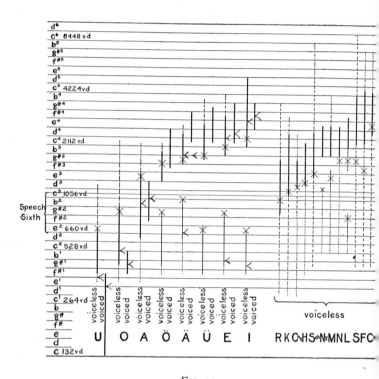

Fig. 13.

The accompanying table reproduces one in which Stumpf presents his most recent data on the location of the formants of unvoiced vocables (Fig. 13). It must be remembered, of course, that these sounds of the German tongue vary somewhat from English and American vocalization. The heavy lines indicate the range of the primary formant for each

of the vowels and consonants investigated. Secondary formants are indicated by lines more lightly drawn. A range
of resonance which is noticeable, though less important, is
recorded by dashes, and a further range, still less important,
by dots. Wherever a characteristic single *pitch* was noted,
it was estimated by Stumpf, and its location is given by a
cross-mark. I have also added to Stumpf's diagram the
formant-regions of the voiced vowels, and upon each I have
indicated by an arrow the characteristic levels of Miller's
analyses.

As regards the sounds assigned to the consonants, the *r*
was lingual, the *k*, soft as in "kö," *ch*$_g$ was the guttural of
"ach." *Sch* was of medium brightness, but *s* sharp and
hissing, and *ch*$_p$ the palatal sound of "ich" or "chi." *Ng,*
m, and *n* were nasal noises—not the diphthong *ng*. Stumpf
remarks that he regards the consonants as vocal noises, both
toneless and colorless, whereas the vowels are vocal clangs,
tonal and colored. He finds the difference analogous to that
of achromatic and chromatic visual sensations with similar
possibilities of transition, such as are found between neutral
grays, grays with some nuance of color-tone, and fully sat-
urated colors. The distinction as it appears in voiceless
vocables shows that some tone attaches to the unvoiced
vowel, less to the consonant. The sharper differentiation of
the formant-regions in vowels points to this distinction, the
consonants showing a more gradual and evenly distributed
resonance from the bass upwards.

In comparison with voiced vocables, whispered vocables,

whether vowel or consonant, show a more even distribution of resonance, extending as Stumpf describes it, like a kind of *tone-dust* continuously or with imperceptibly small intervals over a long tonal range (*125*, 235).

This fact, taken in connection with the apparent difficulty of isolating a formant in perception, indicates that this component of the vocal structure is perhaps never perceptible as an object, but only as an emphasis within a perceptual unit of sound. The very dullness of the sound renders it unsuited to cognitive isolation, and hence our means of determining its character are the indirect methods of Stumpf's and Miller's analyses.

CHAPTER V

NOISE

THE PRODUCTION OF NOISE

To the physicist noise is occasioned by aperiodic vibrations. Among the agitations of the resonating medium there may be some which by reason of their brevity, or of a continuously changing period, fail to set up a pendular-formed motion. Fluctuations of the air having no regular period of frequency are often regarded as the characteristic source of noise. These confused vibrations are supposed to act upon the receptor as jars or shocks to the nerve-endings. However, two things must be taken into account. First, in so far as these jars or shocks are communicated through the air, this medium, being elastic, will naturally tend to respond with a pendular type of wave-motion. Irregularity with respect to the origin, or generator, of a sound does not necessarily occasion an irregular action of the resonator by which the sound is transmitted to the ear. The second point concerns the action of the ear itself.

Some think that the ear has a special organ for noise, this being the vestibular apparatus, or "shake-organ," which is found in lower forms of animal life such as the cœlenterates. This primitive type of ear is but a sack-shaped opening in the side of the body, lined with sensitive cells, and containing hard calcareous particles. When the body is shaken the calcareous particles, or *otoliths,* as they are called, strike against and excite the sensitive cells. From this resemblance to a child's rattle-box arises the name "shake-organ." The human ear contains this organ in the vestibular cavity, where minute otoliths are found embedded in a gelatinous mass resting upon hair-cells. The idea that the stimulation of these cells by shocks and jars is the occasion for noise, as distinct from tone, has not been generally accepted either by anatomists or by psychologists. That lower forms of life possess this organ is, of course, no guarantee that they can hear sounds, although there is some evidence of hearing in fishes that have no cochlear apparatus (*83*). Positive evidence that an animal hears sounds is often difficult to secure because its behavior upon which the inference is based may be the result of a fine sensitivity to intermittent pressures and contacts. The organs of the vestibule, the primary function of which in man seems to be that of equilibration, may serve the fish likewise as receptors of shocks and jars mediated through the water in which it swims, while the fish is no more conscious of a noise or other sound than we are conscious of a direct sense of equilibrium.

It is not improbable that the vestibular apparatus is a receptor of jars and shocks; it may even respond to sound-waves, as G. W. Stewart has suggested in his attempt to explain the localization of sound (*114*); but that it is likewise the normal receptor for noise is an entirely hypothetical, and indeed a gratuitous, assumption. It is much safer to suppose that the physiological seat of hearing is confined to the nerve-endings of the cochlea, which terminate, like other sensory nerves, in the cerebral cortex, whereas the vestibular nerves terminate in the cerebellum, whose function appears to be motor rather than sensory. At least we have reason to believe that the receptor for tone is likewise a receptor for noise. The exact nature of the sound heard must then be bound up in some way with the afferent integrations taking place in the several nerves stimulated by a succession of neural impulses beginning at the receptor and ending in the cortex. Noise, in principle at least, is therefore reducible to the same set of attributes discovered in the perception of tonal and vocal sounds. The physical conditions of the impression would seem to indicate the presence of pitch, brightness, volume, intensity, and duration, just as they do in the case of any other sounds. A study of the primary integration of these attributes as they are observed in a simple perception of noise is therefore our first consideration. We may begin our analysis with a comparison of tone and noise.

THE COMPARISON OF TONE AND NOISE

The connection of tone and noise, as these sounds are commonly heard, is notably intimate, for practically all tones are more or less noisy. Even the "pure" tones obtained with the aid of a refined technique are usually somewhat noisy. On the other hand it would be difficult to find a noise that is not to some extent tonal. Drop a book on a hard-wood floor, and it will make one kind of noise; drop it on a wooden table, and it will make another; for the second will almost certainly be judged "higher" and "brighter." Upon close observation we find that most noises possess a characteristic "height," or "brightness," just as do tones; a series of properly graded bars of wood, if provided with an appropriate resonator, constitutes a musical instrument known as the xylophone.

Another observation also points to the close connection of tone and noise. As we approach the upper and lower limits of audibility, tones seem to pass over into noises, even though their stimuli remain simple pendular-formed vibrations. We have remarked that sounds are audible within the limits of 16 v.d. and 20,000 v.d. As we pass below 150 v.d. or above 4,000 v.d., the sounds take on a decidedly noisy character. In the lower region they are dull rumblings, while in the higher they are piercing squeaks. Indeed, according to the results of certain investigators, the range for these noises extends considerably beyond the above-mentioned limits—downwards as low as 10 v.d. and upwards as high as 60,000 v.d. Certain difficulties of a physiological order furnish a plausible explanation of this change from tone to noise; for towards the limits of audibility the aural mechanism probably loses its capacity to respond in true form. The stimuli are appropriate enough, but at these ranges they seem to shock the organ; so that its action becomes vague and lacking in precision. In consequence, other elements of sound are introduced along with the simple tone, and the total effect is that of a noise. These varied observations upon tone and noise do not, of course, demonstrate that the seat of hearing in man is exclusively in the cochlea, but they point to the conclusion that both noise and tone are perceptual products which may be reduced to the same attributive components.

Let us therefore attempt a psychological analysis of these two percepts in terms of the attributes we have already distinguished. Tone and noise each have pitch, brightness, intensity, duration, and volume. With tones, whether simple or complex, a definite pitch tends to emerge and dominate all subsidiary pitches, causing them to blend with it in an orderly manner according to certain fixed musical principles that will be discussed in subsequent chapters. With noise, on the contrary, a single pitch emerges less definitely, and in any complex perception of noise there may occur many pitch-salients that have no orderly arrangement or interrelationship within the total sound-mass. And hence a perceived tone may be described as an orderly mass of sound, in which a definite pitch-salient is prominent, whereas noises are disorderly sound-masses whose pitch-salients are confused and unmusical in arrangement. This is essentially a musical distinction. Certain sounds can be used for musical purposes: these are tones, complex as a rule, but containing nevertheless a dominant pitch. Other sounds which are unmusical are either vocalic formants or noises. At the level of ordinary perception the line of demarcation is not a sharp one; for some sounds, normally accepted as tones, are, in fact, very noisy, their tonality being largely supplied by the imagination. Instances of this are found in certain tones of the piccolo and of the double-bass; which, when heard in isolation, sound more like noises than like tones. Other sounds that may be used as tones are commonly regarded as noises; as, for instance, those that result from rapping on a piece of wood, from beating upon a drum-head, or from clashing cymbals.

To illustrate a combination of sounds which may be so varied as to produce either a tonal or a noisy effect, let us consider the use of a number of keys on the piano. If we strike a common chord extending over two octaves, employing the notes c-e-g-c¹-e¹-g¹-c², we hear a tonal mass dominated by the c's, particularly the lowest. If, instead of these seven keys, we strike seven others: c-d-e-f-g-a-b, the effect is that of a noise; for no single tone dominates, and the total impression is a confused one. While analysis is still possible to the practised ear, these seven tones, especially when sounded in the lower register of the scale, are unquestionably noisy.

THE PRIMITIVE NATURE OF SOUND

It has often been suggested that the most primitive quality of sound is noise-like, the perception of tones being a result of later evolution. Jaensch (*42*), for instance, ascribes noise to a less refined sensibility than that requisite for tone, and he regards it as probable that man heard noises in the variable regions of pitch before he was able to concentrate upon and distinguish tones of simple frequencies. The vocalic formants, however, represent for Jaensch a stage of evolution midway between noise and tone: they indicate a partial refinement of hearing in the response made to vibrational complexes confined to a certain region within the audible range of sound. His conclusion is that vocalic sounds are specific qualities of the noise-sense, rather than of the tone-sense, as Köhler had maintained (48, 50). We, on the contrary, have attempted to show that vocables are neither the one nor the other. Original sound was doubtless a complex phenomenon. Whether it was regarded as tonal, vocalic, or noisy must have depended then, as it does now, upon the use to which it was put. Differentiation being always made with reference to some compulsory unit-character, it seems highly probable that gross units of sound were apprehended earlier than the more refined units requisite in the perception of tones and vocables. Thus the significance of noise is doubtless more primitive than the significance of speech or of musical sounds. But if we may assume for aboriginal man the same physiological receptivity possessed by civilized man, then the same compulsory integrations which enable us to distinguish the typical patterns of tone, vocal formant, and noise, would be aroused in both cases. All we can say is that civilized man has learned to discriminate certain phenomenal structures which aboriginal man overlooked.

To our way of thinking, the sound of the vocable is a typical perception, distinguishable from the sound of a tone or of a noise mainly with reference to the smooth dullness which characterizes the vocal formants. Tones depend upon a pitch-salient for their existence, while vocalic formants and noises lack a definite salient. In the case of noise there may be several salients, but if so their order is confused and irregular in the total sound-mass. But noise may also be phenomenally simple without this confusion or irregularity.

At the lower levels of pitch all sounds are noisy and dull; the smooth effect of a true vocalic formant is here lacking, for the same reason that a definite pitch fails to emerge. The region in question may be said to suggest some of the characteristic features of man's primitive consciousness of sound as conceived by Jaensch, where a regional pitch is supposed to have been heard whose effect was vague and indefinite. The formants of the vocables which occur at higher frequencies would likewise be lost, were it not for our ability to discriminate the complex perceptual patterns in which they occur. At our present stage of culture we are able to detect uniformities in a gradation from noise through vowel to tone as a more primitive mind certainly could not. Among imbeciles a similar disability is manifest. A discriminative ear for tones may require even greater perceptual ability than does an ear for speech, because in music we must be able to order the pitch-salients of sound with respect to their volumic proportions in intervals that presuppose a cultural tradition, both of tonal impression and of expression.

At the higher levels of pitch, sounds also possess these primitive characteristics—in as much as they are both noisy and lacking in the roundness of effect needed for the production of vocal formants. Yet the audible range of unmusical high tones is much greater than that of the corresponding unmusical low tones, and the larger number of discriminable pitches and intensities of this range materially aids in the perceptual discrimination of high-pitched sounds, because the *distance* between different pitches is apparent even where a musical *interval* is not.

THE SIMPLE PHENOMENA OF NOISE

We have thus far described the perceptual characteristics of a complex sound, and now must consider the primary integration of noise as compared with the primary integrations of the "pure" tone and the vocal formant. Although the ordinary noises of everyday experience are no purer than the tones and vocal utterances we hear, it is possible to select conditions under which a greatly simplified sound of this type may be produced. The most important of these conditions is *time*. While tones are dependent upon a certain vibrational frequency, a noise can be produced by a single non-recurrent pulsation.

The siren is a useful instrument in the study of simple noise, since the individual vibrations produced by the air as it is allowed to pass through the holes of a revolving disk can be controlled both as regards their number and their frequency. The experimental results of Abraham (2) again furnish us with our principal data. The sound produced by a small number of openings in the siren-disk—the rest being closed—is still tonal, though decidedly noisy. As the number of openings is reduced, the noise becomes more apparent, while the tone is less pronounced, although a tone is still discernible if only two holes are left open. When there is but one opening, however, only a noise is heard.

The question arises whether this noise accompanying brief tones, and characterizing even a single pulsation, is not really a product of secondary resonance, either in the air or in the ear. In order to test this, Abraham made a number of very carefully controlled experiments. He found that every brief tone of the siren was accompanied by a clap-like noise of which the pitch, though somewhat ambiguous, appeared to be about an octave lower than the tone which it accompanied. The noise was not altered in any way by changing the position and distance of the generator with reference to the walls of different rooms that might reflect and therefore resonate the sound. He found, too, that the noise remained when the tonal accompaniment had been completely eliminated by interference. The pitch of the noise varied neither with the frequency nor with the resonance, but with the absolute time of the pulsation. If the time was less than $1/150$ of a second, the character of the noise was altered, and, instead of the clap-like sound, a blowing effect was produced. This alteration of the noise could be brought about with a constant vibrational frequency by employing holes differently spaced in different circles of the disk, so that the frequency of the pulsations remained constant, while the time in which the air was allowed to pass through the holes varied from a shorter period at the periphery to a longer period at the centre.

No difference was found between the noises produced when the pulsations came from regularly arranged openings, from irregularly arranged openings, or from a single opening in the disk. In all cases the threshold for the clap-noise was approximately 1/150 of a second. A succession of these clap-like sounds is described by Abraham as a "rattle," and apparently a "clap" can be heard if any disturbance of a tone occasions the grouping of a certain number of vibrations into a single pulsation, provided that the pulsation acts for less than 1/150 of a second. By gradually decreasing the speed of rotation of the disk, so as to lengthen the periods of pulsation, the noise became first a sound like "fft," then a blowing noise whose pitch seemed to be raised several octaves. As the speed was still further decreased, the new pitch was again lowered, until at length the sound disappeared altogether. If, on the other hand, the period between pulsations was shortened by speeding up the rotation of the disk, the noise increased in brightness, and it also appeared to become vocalic.

A simple noise may be either "rough" or "sharp" according to its degree of dullness or brightness. Yet "roughness" is something more than the dullness of a vocal formant, and sharpness something more than the brightness of a pure tone. It is possible that we have here a new attribute by which to distinguish the simple noise from the vocable and the tone, but it is more likely that both "roughness" and "sharpness" are integrative characteristics of the noise to which pitch, brightness, duration, intensity, and volume all contribute.

Further investigation is necessary before the precise details of these phenomenal structures can be known. We may draw certain inferences, however, from Abraham's results. Let the tonal vibration be represented by a sine-curve, and assume, with Abraham, that under certain conditions the curve becomes asymmetrical by a lengthening either of its crest or of its trough. In the first case of asymmetry the sound becomes dull, and finally rough and noisy. In the second case it becomes brighter, and finally sharp and hissing. Thus the two types of noise would be conditioned by the two kinds of asymmetry. The vocalic series from *a* downwards to *u* would accompany the first degrees of dullness, while the higher vowels, *e* and *i,* and the sibilants would accompany increased brightness.

It must be admitted that these inferences are tentative, for we do not yet know all the facts, including the place in the vocalic series where the change from symmetry to asymmetry takes place. Furthermore, whenever there is a mixture of frequencies, the higher partials may lend sharpness to the sound, despite the fact that its fundamental remains dull and rough. In the appropriate regions of the vocal formants, a sound thus modified by interruption will resemble a vowel whenever the formation of its structure is favored by a resonance which smoothes its roughness and reduces its sharpness; for this appears to be the manner in which a sound is transformed into a true vocal formant. A pitch-salient is introduced only when the sound-pulse recurs regularly at a given frequency; but the combination of a definite frequency with the asymmetry of wave-length occasioned by interruptions also gives rise to partial tones. It is therefore improbable that an uninterrupted "pure" tone can ever be noisy, though noise frequently accompanies tonal clangs. The primary integrations of pitch with brightness or dullness, which characterize in turn the noise, the vocable, and the tone, may be described as a transition from roughness or sharpness, through smooth-dullness, to saliency. It should be remembered, however, that attention plays an important part in the formation of these structures, and in the definition of their perceptual contours. The physical condition under which a simple noise is integrated is a primary non-recurrent wave-length, while the condition for a tone is a regularly recurrent frequency. Between the two lie the conditions for the vocal formant, tone-like because it depends upon a certain frequency, noise-like because its wave-lengths are irregular. The vocable passes into a tone as its period of frequency is emphasized; for instance, in singing a vowel in harmony with the region of its formant, the disturbance which normally lengthens one part of the wave, causing a reduction of its brightness, is in large measure overcome because the formant now resonates as the partial of a vocal clang; accordingly the vowel is sung as a tone. The sound loses its vocality, however, in the measure in which it becomes tonal, as Abraham likewise has shown in another investigation (*3*, 134). On the other hand, the vocable passes into a noise whenever the disturbance is sufficient to destroy the regional resonance upon which the formant is based.

Hence the attributive integration of pitch and brightness supplies the essential characteristics of these elemental variations. Yet we must not suppose that the attributes of volume, intensity, and duration are negligible. It is only because we know less of the parts they play that we have omitted them so largely from our consideration. The clap-like noise accompanies a tone of brief duration, yet its condition is not brevity *per se,* but an undue lengthening of the wave, introduced either by the grouping of a small number of pulsations, so that together they have the combined effect of one pulsation, or by shortening the pause between single pulsations, so that the trough of a single wave—its period of condensation—is shorter than the crest—its period of rarefaction. Both these last-named conditions provide for an increase in the duration of a special component which appears to determine the variation of brightness; but conditions are likewise given which might also involve modifications of volume and intensity. We must await further investigation of the phenomenology of tone, vocable, and noise under varied conditions of production, before we can attempt to define their several integrative patterns with assurance. Meanwhile the description of the three types of sound as they ap-

pear to us in the more complicated forms of perception will
serve as an indication of the simpler and more elemental
structures that underlie their differentiation.

SUMMARY

The elemental nature of sound is described in terms of five
attributes: pitch, brightness, intensity, duration, and volume.
There are also three definite kinds of auditory perception:
tonal, vocal, and noisy. Each of these percepts may be
analyzed with reference to the five attributes which are
aspects of each and all of them. Accordingly we have no
present need for assuming an elemental tonality, vocality,
or noisiness. The transition from tone through vowel to
noise is not unlike the transitions noticed in the color-series.
Yet an important difference is also manifest; for whereas
in the color-series we pass through "intermediates" from one
"quality" of experience to another totally different from it,
in sound the transition is marked only by a shifting emphasis
among the attributes, all of which are present in each several
kind of sound. A tonal percept may be at once noisy and
vocalic. Similarly, a vocable may be both tonal and noisy,
and a noise may have tonality and vocality. But a perceived red can not at the same time be green. All three
perceptual integrations of sound possess the same attributive
aspects in varying degrees of distinctness: pitch, brightness,
intensity, duration, and volume. The clear emergence of a
pitch-salient, commanding and subordinating all other sali-

ents that may be present in the sound-mass, is the characteristic of tone.   Lack of such a predominance, in a complex
sound, through lack of order and a regular subordination of
the salients of pitch, together with a rough dullness or a sharp
brightness occasioned by an asymmetry of the sound-wave, is

Fig. 14.

the characteristic of noise.   The vocalic formants seem to
occupy a middle ground, for while a certain regularity is
indicated among the several components of a vocable, the
formants lack saliency of pitch, although they appear to
possess the smoothness of a tone rather than the rough-

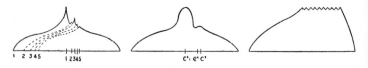

Fig. 15.   The first design represents a musical clang with its fundamental and first four overtones.   The second represents the vowel
i ("eat") with formants in the regions of c^2, e^4, and c^5.   The third
represents a noise produced by a mixture of non-harmonic tones.

ness of a noise.   Of the formants which characterize the different vowels, each has a definite place in the scale, and we
have suggested that their integration differs from that of a
tone by reason of a qualitative change from brightness towards dullness.   As Abraham has shown, this suggestion is

something more than a conjecture; since a difference of brightness may be noted in tones of the same frequency, provided that a modification in wave-length takes place without a corresponding alteration in frequency. In the "pure" tone a single pitch emerges to dominate and give point to the sound-mass. In the formant we seem to experience a more globular effect, smoothly rounded, and without a definite salient. A simple noise has neither of these characteristics, but instead a rough dullness or a sharp brightness whose regional pitch depends upon the primary wave-length of a single vibration or of an isolated group of vibrations.

The transition from a tone of salient pitch through one of a lesser degree of brightness to the smooth and non-salient formant, and thence to the rough dullness of a simple noise, is graphically described in Figure 14. In addition, the dimensions of sound which result from certain complexes of vibration as they unite to produce a musical clang, a vowel, and a noise, are set forth in Figure 15. The construction of both these figures will be clear from what has gone before.

CHAPTER VI

TONAL FUSION

PITCH-ORDER

WE have already noted that by varying the relative positions of the pitch-salients within a single perceived mass of sound we could produce either a tone or a noise. As an illustration we cited the case of seven tones, in the order of the common chord of two octaves, which when sounded together give a tonal result; this we compared with seven adjacent tones of the scale, which, when simultaneously sounded, give the effect of a noise. In the first case there is a certain orderly arrangement of the components, not found in the second case. What, then, is this order, and how does it arise?

Order is responsible for what we may term the *character* of tones, and is, accordingly, the foundation for the language, or mode of auditory communication, that we call music. There are many theories as to its origin, but, before considering some of them, we should familiarize ourselves with certain underlying facts.

Pitch, we have seen, signifies a graduated series of sounds rising from low to high. The lowest and the highest are at the limits of audibility, these limits being attained at vibrational frequencies approximating 16 and 20,000 in the sec-

ond.   Between them there stretches a graded series of
sounds, which, if relatively simple and tonal, yet comprise
a large number of just noticeable differences of pitch-brightness.   In considering the order of tones, however, we are
concerned not merely with just noticeable, or *liminal,* differences of pitch, but more especially with certain *supra-liminal*
differences involving the integration of pitch with other attributes, volume in particular.   These supra-liminal differences are more obvious in everyday life, and follow a different law.   The striking thing in the discrimination of tones,
as evinced by the musical usage of different peoples, is not
the ability to discriminate just noticeable differences of pitch-
brightness with equal increments of vibrational frequency,
but the ability of the ear to detect equalities of *interval,*
based upon like *ratios* of vibrational frequency.

INTERVAL

The sense of interval is at the foundation of all music;
but interval is something more than a simple difference of
pitch.   Difference of pitch can be detected when there occurs but a fractional change in the vibrational frequency of
the sound.   Yet such a difference does not constitute an
interval; for we still hear the same tone, slightly modified as
higher or lower.   That which affects a tone, making it appear different *as a tone*, is a change of vibrational frequency
sufficient to let us discriminate a change in *volume*.   Difference of volume follows another law than that of differ-

ence of pitch. While equal increments of one vibration or less may occasion difference of pitch, only by larger increments which traverse several observably different pitches can an interval be detected between two tones. Experimental results bearing upon the distinction of tonal *distances* and tonal *intervals* are conflicting,[1] but observation seems to show that *distance* is primarily a function of the number of ascending or descending pitches involved in the step taken, whereas an interval requires not only a minimal distance, but also a certain volumic proportion of the tones involved. Hence the distance between two tones of an interval like the octave may appear less at the lower levels of the scale than it does at the higher levels, because fewer discriminable pitches are traversed in the former case than in the latter. Since equal intervals depend upon equal proportions rather than upon the summation of increments of difference, we shall find a volumic difference more significant in the determination of a given interval than the number of discriminable pitches included in the interval, because this number must vary if the same interval is repeated in different regions of the scale.

With the results at hand, it appears that the differential *limen* for volume is attained when the vibrational frequency is increased by from two to four one-hundredths of the standard frequency. As compared with this *limen*, the semitone interval of our scale is produced by an increment of

[1] Cf. *122*, I. 57 ff.; also a recent statement of the problem of *distance*, together with some experiments, by Pratt (*91*).

approximately .06. Thus it appears that the steps by which intervals are distinguished, are determined by our ability to discriminate volumes. The employment of small intervals is therefore limited, for although the smallest interval used in our music is considerably larger than a bare difference of volume, it would be impossible to use intervals much smaller without approaching the region of liminal uncertainty.

The most prominent interval, and the one that appears in the music of all races, is called an *octave*, because the interval happens to embrace eight tones in one of the important scales of our music. But the name is otherwise fortuitous, since the interval may be divided in many other ways. The essential physical condition for the octave is an interval of two tones having vibration-numbers in the ratio of one to two—a convenient definition, although our sense of hearing sometimes deviates from the mathematical accuracy of the ratio. For general purposes, however, we may regard the octave as implying sound-vibrations in the ratio of one to two. Thus, tones of 100 v.d. and 200 v.d. are in this relationship, as are tones of 256 v.d. and 512 v.d., or indeed, any other tones in which the ratio 1:2 is closely approximated.

FUSION

When two tones in the relation of an octave are heard together, they seem to blend into one uniform impression of sound. In a word, they *fuse*—a very significant fact, there being many fusions among tones, but none quite so important as this. Here are two tones, widely separate in pitch, which nevertheless blend into a uniform whole. Let one of the two be but slightly shifted upwards or downwards in the scale, and all uniformity ceases; the effect of fusion is lost, and in its place we have a discordant clang. But not only do these two tones fuse when sounded together; they also appear alike when sounded successively. They have a similar *tonality*.

The ascription of tonality carries us back to our discussion of the attributes of tones, where the provisional decision was reached that tonality is not a sixth attribute, although many so regard it. But we shall not follow them, because tonality seems to us better systematized as a product of the musical setting than as a specific aspect of a single sound. According to our view, the octave is a *relation*—a meaning of one tone for another. The two tones of an octave *mean* the same thing for musical purposes; and hence they have a like tonal "character," a word we use advisedly, because it suggests "functional" activity. A tone taken by itself is an integration of various attributes, but it is only for musical purposes, and with regard for musical thought and practice, that it enters into perceptual union with another tone to become a member of an octave. To the unmusical this feature has less significance than it has to the musical; yet the law of the octave rests upon no arbitrary, man-made convention; musical thought has its incontrovertible logic, its foundations being laid by the conditions, physical and physiological, under which we hear.

The octave brings order into the series of tones; for at stated intervals each tone finds its like, both above and below it in the series. We might graphically represent the whole series as a spiral extending from low to high. A vertical line on the spiral, drawn parallel to the axis, would then mark a series of octaves. But the figure would only partially represent the phenomena, for the octave is not the only instance of fusion and relationship. Other simple ratios—2:3, 3:4, 4:5, etc.—indicate lesser degrees of fusion, and a comparative likeness of their constituent tones. The limit of fusion appears to be reached when we employ as ratio-numbers, representing the vibrational frequencies, any two adjacent digits above 7. Thus the essential numbers in the scheme of ratios whose tones will fuse are 2, 3, 5, and possibly 7. The obvious simplicity of these four numbers is somewhat complicated by multiplications introduced in accordance with the law of the octave that, for musical purposes, a number and its double are identical. If a tone corresponding to 256 v.d. is represented by the number 2, a one of 512 v.d. would be represented by the number 4; but since the second tone is the octave of the first, it is musically identical with it. Accordingly, the number 4 equals the number 2; as do the numbers 8, 16, 32, etc. Similarly the number 3 may be interchanged with 6, 12, 24, etc., and the number 5 with 10, 20, 40, etc. An interchange of this nature is usually made for the purpose of reducing an interval greater than an octave to one that falls within the octave. For instance, the interval of the twelfth, 2:6 is analogous to the interval of the fifth, 2:3, which has a similar fusional effect, though enhanced by the proximity of the two tones. In the case of reversed ratios, however, like the fourth, 3:4, the interval has a different musical effect from the fifth, 2:3, even though the upper tone of the fourth (4) is in a sense identical with the lower tone of the fifth (2). This indicates the importance attaching to the position of the tones within a standard interval, which is nothing other than a division or segment of an octave.

It is not easy to fix the precise limits of fusion, because practice and familiarity tend to dispose us more favorably to the acceptance of intervals which at first may be rejected as non-fusing. For our present purpose it will be enough to note that tones of small ratio-numbers which are neither too close together nor too far apart in pitch will evince some degree of fusion when sounded together, and likewise some degree of relatedness when sounded in succession. As for the many intervals of tones that do not fuse, the musical relationship of which is therefore less evident, it yet remains true that like ratios tend to be regarded as like intervals *in any region of the scale*. Even a quite dissonant interval such as 10:11 can, with practice, be recognized upon whatever tones it may be heard. As we presently shall see, there are good grounds for supposing that this power of judging equality of interval upon the basis of equally proportioned segments is more primitive, and more fundamental to music, than is the ability to perceive fusion, or to detect the relation of one tone to another when they are heard successively.

The blending or fusing of tones depends upon an orderly arrangement of the components, and the order implies certain simple ratios of the vibrating frequencies. The most conspicuous of these ratios are 1:2, 2:3, 3:4, 3:5, 4:5, and 5:6. Other ratios, such as 4:7, 5:7, 5:8, 6:7, 7:8, 7:9, 7:10, 7:12, 8:9, 8:15, 15:16, are commonly regarded as non-fusing. Yet they manifest varying degrees of failure, just as the ratios first mentioned fuse with varying degrees of success. Furthermore, the musical relationship implied in some of the intervals of the second list is very close, despite their lack of fusion when sounded together. Indeed, relation and fusion follow different functional principles. The former is a product of musical thought and expression, and is perceptually complex, whereas fusion is a primary integration, compulsory in its effect, and may be referred with some confidence to the physiological mechanism of aural receptivity.

Of the numerous theories advanced to explain fusion, we shall examine six, for the inquiry can furnish us many suggestions regarding the significant part that fusion plays in the act of hearing. The first five of these theories interpret fusion as an inherent result of conscious, subconscious, or physiological aggregation; that is to say, they reduce fusion to an inherent structural effect, produced by the blending of components that have some natural or inherent affinity.

After pointing out the difficulties involved in each of these various attempts, we shall advance a *genetic* or *functional* theory that seeks to explain the structure of fusion by reference to biological selection.

THE HELMHOLTZIAN THEORY OF FUSION

Let us first consider the theory advanced by Helmholtz (*35*, 182 ff.), the key to whose interpretation of hearing was the physical fact of resonance. His principle not only supplied him with his theory of the auditory mechanism, but also led to his explanation of tonal perception. With him, the chief emphasis is laid upon sympathetic resonance as the occasion for partial vibrations. Now it has been noted above that a vibrating body tends to fluctuate in numerous periods other than that of its fundamental. Thus the most emphatic partial vibrations of a freely swinging generator are those twice, three times, four times, etc., its fundamental period. A string vibrating 200 times a second tends to vibrate also in halves, at the rate of 400 v.d.; in thirds, at the rate of 600 v.d.; in fourths, at the rate of 800 v.d.; etc.

Now suppose we have two strings, each set in vibration, one of them stretched or tuned at 200 v.d., the other at 300 v.d. If these are struck together, the two resulting tones will represent the fused interval of the fifth, the ratio of their vibrating periods being 2:3. But if, in addition to their fundamentals, we also take account of their overtones, the combination becomes much more complicated. The tone of

200 v.d. has overtones corresponding to 400 v.d., 600 v.d., 800 v.d., 1,000 v.d., 1,200 v.d., etc.; while the tone of 300 v.d. has overtones corresponding to 600 v.d., 900 v.d., 1,200 v.d., etc. When these two vibrational patterns are superimposed, we note, first, that two of the overtones of 200 v.d. coincide with two of those mentioned for 300 v.d.; namely, the partials 600 v.d. and 1,200 v.d. Secondly, we note that the non-coinciding partials are clearly separated from each other by a minimum of 100 v.d. But suppose we had selected as our original tones those of 200 v.d. and 275 v.d.; then the overtones of the latter would be 550 v.d., 825 v.d., 1,100 v.d., etc. In this case there is no coincidence, nor are the two sets of partials always widely separated; for instance, the partial 825 v.d. is but 25 vibrations removed from the 800 partial of the tone 200 v.d. But what happens when two tones as close as these are sounded together? Beats are heard, because a reciprocal interference is set up between them, and the fluctuations of intensity that result are disturbing to the smoothness and agreeableness of the combined sound. According to Helmholtz, tones of this sort do not fuse, but stand apart on account of this jarring interference. In the former case, however, the tones fuse, both because there is no such interference, and also because they are bound together by coincident partials.

Although at first view the theory of Helmholtz seems very plausible, it has not withstood the criticism to which it has been subjected. As regards the positive effect of coincident partials we are at a loss to understand why this should create a fusion, when all that we have a right to expect is a possible increase in the intensity of the coinciding components. When identical tones are sounded together a unison is effected, the resultant sound being somewhat louder than the sound of either of the components. If there is no difference in phase to reckon with (for that would lead to interference), this is the only effect produced. Accordingly, this observation furnishes no ground for supposing that two tones must fuse when certain of their partials coincide. As regards the more important negative effect of *beating* partials, objection is raised because with the aid of appropriate interference-tubes it is quite possible to eliminate these partials, and when this has been acomplished we might expect that hitherto non-fusing tones should fuse; but such is not the case. Even with pure tones, the same combinations fuse, while others fail to do so.

Many other arguments have been brought against this theory, but in view of the crucial significance of those mentioned, we shall not pursue the matter further—not that the presence of beats or of coincident overtones can have no effect whatever upon fusion, but that a theory which rests entirely upon these facts is inadequate.

KRÜGER'S THEORY OF FUSION

A second theory that in recent years has gained some assent is analogous to that of Helmholtz in its attempt to explain fusion by reference to coincidence and interference within the sound-mass. This theory, advanced by Felix Krüger (52), looks to the difference-tones, rather than to the overtones, for its foundation. We have observed that when two tones are sounded together, a third tone may be produced, corresponding in pitch to the difference in vibrational frequency of the two generators. Other difference-tones that correspond with differences between difference-tones and the fundamental are also heard. Taking our instance of the fifth: in addition to the two combining tones, 200 v.d. and 300 v.d., an additional tone may be heard that corresponds to their difference; namely, a tone of 100 v.d. According to Krüger, there are five difference-tones whose derivation is indicated by a process of successive subtraction. Beginning with the difference of the two fundamentals, we proceed by successively subtracting the smallest vibration-numbers remaining, which include both the lower fundamental tone and the difference-tones as they are derived. In the case of the fifth, these differences are identical, since the difference of 2 and 3, the ratio-numbers of the fundamentals, and the difference of 1 and 2, the ratio-numbers of the first difference-tone and the lower fundamental, are alike 1. All unite, therefore, to produce a single difference-tone. There are no beats of interference, and the whole effect is smooth and agreeable. But let us see what happens with our example of a dissonance—the tones 200 v.d. and 275 v.d. As derived by Krüger, the first difference-tone would be 75 v.d.; the second, 125 v.d.; the third, 50 v.d.; the fourth, 25 v.d.; and the fifth, 25 v.d. Beats would occur between all pairs of these tones excepting the last; for roughness can be detected with differences greater than 125 v.d., and the disturbance would be quite noticeable with differences of 25 and 50 v.d.

Assuming that difference-tones originate within the organism, rather than in the generator, as do the partial tones, this theory gains a certain advantage over that of Helmholtz, inasmuch as the difference-tones would be deemed always present in any combination of tones, whereas partials are fortuitous and may be lacking altogether. Difference-tones also allow of more varied and more serious interferences than are found among the partials. Yet as a complete explanation of fusion Krüger's theory is no more satisfactory than Helmholtz's. In the first place, difference-tones are usually weak to the point of inaudibility, being in this regard far less noticeable than the partials. Again, the theory has been successfully attacked on experimental grounds. Krüger's five orders of difference-tones, and his method of their derivation, are both questionable. In a careful experimental investigation of combination-tones— both difference- and summation-tones—Carl Stumpf (*116*) found several orders of difference-tones, but his results did not confirm Krüger's derivation. Stumpf also observed (*119*), that the interval we have used as an example of a dissonance would not always have the effect attributed to it by Krüger's analysis. This interval of 200 v.d. and 275 v.d. has the ratio of 8:11. If the tones were of 800 v.d. and 1,100 v.d., the difference-tones, as Krüger computes them, would be 300 v.d., 500 v.d., 200 v.d., and 100 v.d.

Here one finds no serious interference, the closest interval being that of a fifth. Should we not, then, in this case expect a perfect consonance, instead of the well-established dissonance of this ratio? Thus each of these theories derived from the interaction of components within the sound-mass appears inadequate as an explanation of the peculiar blending of tones we call fusion; the theory of Helmholtz because the partial tones may be altogether lacking; and the theory of Krüger because his derivation of difference-tones is incorrect, and also because in a critical instance his hypothesis breaks down.

LIPPS' THEORY OF FUSION

A third theory of tonal fusion, advanced by Theodor Lipps (58), involves the assumption of "micro-psychic" rhythms; that is to say, rhythmical processes that are unconscious, but which contribute to a resultant fusion of tones when the rhythms of the sound-vibrations are compatible, and to a lack of fusion when they are incompatible. With tones of 100 v.d. and 200 v.d., micro-psychic rhythms of 100 and 200, respectively, are assumed. As these coalesce into one rhythmical pattern, the more rapid rhythm adjusts itself to the less rapid, so that every second beat of the 200 fluctuations coincides with one of the successive beats of the 100 fluctuations. With the interval of a fifth, whose ratio is 2:3, we should have a coincidence of every third beat of the higher sound with every second beat of the lower.

As the numbers representing the ratio of the interval increase, the coincidences in the period decrease; hence lesser degrees of fusion result. Of the numerous objections to this theory, the chief one, perhaps, is the removal of the basis of explanation from the realm of the conscious to that of the unconscious. The rhythm of a tone, as indicated by its vibrational frequency, is a thing of which we are not at all aware. Consequently, the hypothesis amounts to little more than the tautology that simple ratios of vibration fuse because they are simple, while complex ratios fail to fuse because they are complex. But more specific objections can be brought forward even if we were ready to accept Lipps' principle. In order to be effective, these rhythmical coincidences would require that tones should continue throughout their periods. Yet, as Krüger has pointed out, fusion is not noticeably affected if the tones are interrupted and allowed but brief and incomplete periods of duration. Furthermore, some difference of phase is almost always present in any two sounding bodies, which means that the vibration of one tone will usually start in advance of the other. When this is the case it can not but seriously interfere with rhythmical coincidences; yet a difference of phase appears to have no effect whatever upon the fusion of two tones.

STUMPF'S THEORY OF FUSION

Stumpf attempts to solve the difficulty by attributing an elemental significance to fusion (*122*, II, 184 ff., *121*).

Fusion is an *immanent relation,* giving rise to a uniformity that in itself can not be further analyzed. The uniformity of fusion is quite different from the union of tones in a chord, since the chord can be analyzed into its constituent tones. With the octave, for instance, we may attend to the lower or to the upper tone; but only when we do this is our attention called to the fusion that existed, and that continues to exist even after the analysis ends. With simultaneous tones that do not fuse, analysis is easier, for it becomes apparent that the tones stand apart, and that a fusional uniformity never existed between them. In this way Stumpf demonstrates that fusion and difficulty of analysis are not the same. Quite apart from analysis, fusion is an imminent balance or relationship of the fusing tones. It is not identical with *concordance;* for the latter is a musical term, and its application a matter of musical thought and practice. The blending of tones is something inherent in themselves; no special act of the mind is required to establish it.

Having concluded that fusion is an ultimate element of experience, Stumpf leaves us with no further explanation than a vague reference to a "specific synergy" of nervous function. Even though we were to accept his further assumption of "relational elements" in consciousness, no explanation is offered for the fact that tones whose vibrational ratio is 2:3 *are* related, and capable of fusion, while tones whose vibrational ratio is 8:11 are unrelated, and incapable of fusion.

WATT'S THEORY OF FUSION

Various writers have attempted to preserve the merits, and to avoid the difficulties, of the preceding theories. We shall confine ourselves to two hypotheses on the subject that are now current. The first has been recently stated quite fully by H. J. Watt (*134,* 53 ff.). Watt explains fusion as a *balance* among the tonal masses of the fusing components. When two tones are sounded together, they form a combined sound-mass the most conspicuous attributes of which are pitch and volume. The range of tones from low to high is an orderly series of ascending pitch, attended by an orderly series of decreasing volume. Thus the volume of a low tone must contain within its bulk the volume of every higher tone sounded along with it. The sound-mass of two tones finds the higher tone contained within the volume of the lower at a point determined by the "order" of its pitch and the spread of its volume.

The argument in favor of this hypothesis runs somewhat as follows: Two tones when sounded together interpenetrate each other; they never really stand apart, even though they may not fuse. What, then, causes some tones to blend? Evidently not their pitches, for these remain distinct. Of the remaining attributes, volume is clearly the most significant as a basis for coalescence. Since volume decreases as pitch rises, it follows that the higher of the two tones will have the smaller volume. Assuming the pitch-salient to occupy a central position within the volume of a tone, and with due regard for the ordinal position of each pitch in the tonal series, Watt infers that the higher tone always falls within the upper portion of the volume of the lower tone. This leads him to believe that the upper limiting orders of the two volumes always coincide. That is, when any two tones are heard together, their volumes will combine in such a way that the spread of each mass towards the upper region of sound will reach the same point of extension indicated by the highest ordinal of the pitch-series. Accordingly, all tones, of whatever pitch, have the same upper limit of volumic spread. This particular emplacement of any tone within another of a lower pitch may be determined at once by the ratio of the vibrational frequencies of the two tones.

Take the octave, for instance. The volume of the upper tone is assumed to be just half the volume of the lower tone; therefore the upper tone falls exactly within the upper half of the lower tone-mass. Since the upper limits of the tones coincide, the lower limit of the upper tone's volume must fall together with the pitch-salient of the lower tone, while the pitch-salient of the upper tone will lie midway between the pitch of the lower tone and the upper coinciding limits of volume. This balance, or symmetry, with respect to the predominating salients and the volumic spread of the two tones constitutes their fusional effect. With other fusing intervals the patterns are more complicated, but the upper limits of volume always coincide, while the pitch-salient and the lower limit of volumic spread belonging to the upper tone are always placed at simple divisions of the tone-mass whose outline is indicated by the total volumic spread of the lower tone. Balance, or symmetry, within the sound-mass is thus the cause of fusion. These data are graphically represented in the accompanying figure for the octave, fifth, and fourth (Fig. 16).

Watt supports his ingenious explanation by referring to an analogous response on the part of the basilar membrane. As we have noted in describing his physiological hypothesis (p. 44), portions of this membrane are supposed to be set in action to correspond with the volume of the tone. Thus the pattern of coincident masses of sound has its analog in the agitations of the basilar membrane. Now the membrane is always affected from the base of the cochlea inwards. Therefore a point adjacent to the oval window marks the uniform upper limit of volume for every tone. When the volume is great, and the tone low, a large portion of the membrane is involved, the action that determines the pitch-salient being well towards the middle of the membranous band that stretches to the apex of the cochlea. When the

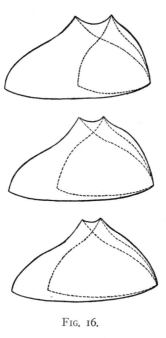

FIG. 16.

volume is small, and the tone high, only a small portion of the membrane responds, and the physiological correlate of the pitch-salient, being central to the part affected, will be found near the oval window.

Despite the ingenuity with which Watt's hypothesis is constructed, it is manifestly unacceptable if the inferences we have drawn from Rich's work on the threshold of volume are correct. Referring to Figure 10 we see that the volume of a tone can not be half the volume of another tone which is an octave lower in the scale. In the middle range of the scale, between 250 v.d. and 1,000 v.d., each octave should contain approximately the same number of discriminable steps or intervals, some twenty-four to thirty in number. Above and below this region discriminability is less accurate. Although the higher tone may perhaps be said to fall entirely within the volume of the lower tone, as Watt contends, it can not occupy just half the volume, and thus determine the "balance" upon which Watt bases the phenomenon of the octave-fusion. We have at present no means of measuring the precise distance of musical intervals, and the scale of our graphic representation in this respect is arbitrary; but at least we can see that if Weber's law that sensory differences are based upon proportional increments of stimulation holds for volume, upon which the sense of interval seems to rest, the volumes of successive octaves are not *halved* as Watt assumes they must be; and if the "predominant orders" of two tones in octave relationship are not so placed as to establish the coincidences which Watt's hypothesis demands, it is obvious that lesser degrees of fusion can not be founded upon this peculiar conception of symmetry and balance.

We must therefore conclude that balance and symmetry are inadequate to explain fusional effects, although Watt has undoubtedly contributed to the theory of fusion by calling attention to the coalescence of volumes when two or more tones are sounded together. Whatever occasions the blending of pitches, we have at least a definite and useful envisagement of the phenomenon in Watt's assumption that the volume of the lowest tone completely encompasses the volumes of all higher tones. But let us remember that it is an assumption; for at present we have no evidence that the volumes of different tones must overlap in this manner.

THE HARMONIC THEORY OF FUSION

The last type of theory to be mentioned concerns both the racial and the habitual selection of certain components of sound for integration, without reference to immanent "relations" or "balances" such as Stumpf and Watt have in mind. The present theory also views the problem in a different way from that which obviously inspired Helmholtz and Krüger in their attempts to solve it. The general idea of the harmonic theory is that certain integrations of pitch are effected by ingrained dispositional tendencies, both innate and acquired; and the question which confronts us is why certain intervals should lend themselves to these dispositional trends while others do not. The answer is given in the harmonic series of the partial tones (*74, 78*).

It is significant that the fusing intervals should be the very ones that are most conspicuous among the partials of a clang. Although relatively pure tones are occasionally produced, most of the sounds we hear are complex masses of partial tones. Within the full resonance of a clang containing sixteen partials (and many more are often distinguishable), the octave occurs no fewer than eight times; the fifth and the fourth occur four times; the major third and major sixth, three times; the minor third and minor sixth, twice. Other intervals that fuse less, or not at all, are also present, but *no dissonant interval occurs more than once.* If frequency of association contributed to one's ability to fuse tones, this would seem to be an obvious source of habituation. But according to Stumpf (*122,* II, 208 f.) such is not the case, and he therefore rejects this explanation; certain recent experiments, however, suggest a reconsideration of the matter. Moore (*73*), Valentine (*129*), and others have found that as one grows used to certain dissonances they come to seem less strange and more uniform than at first. Moore, in particular, has demonstrated improvement after practice with respect to the consonance of the sevenths. The history of music also gives evidence of such adaptations. The octave passed for the most beautiful consonance among the ancient Greeks; during the Middle Ages the fifth achieved a commanding position; and only gradually thereafter was the third admitted to be consonant, though it is now accepted without question.

But apart from individual habit, which, of course, would vary greatly with experience, does not the frequent occurrence of the most striking fusions—the octave, the fifth, and the fourth—among the partial tones, warrant the assumption that adjustments of the ear to frequent combinations of tone can, in some measure, be selected through inheritance? [1] In the first place, the facility with which the dominant partials blend would make it easier to distinguish the odd partials from the accompanying noises that often serve to characterize the object generating the sound. The *timbre* of a sound consists in these partial-tone and noise-like constituents. Thus discrimination of timbre may well be regarded as possessing a "survival value," because an ear capable of fusing recurrent intervals would be better able to exercise a fine discrimination among the non-fusing and less frequent constituents of a sound which determine its practical significance. A ready ability to perceive and set aside the structural uniformities of the more common intervals would leave the mind free, as it were, to discover any uncommon components, whether tonal, vocal, or noisy, which might chance to be present in the sound-complex. This theory of instinctive disposition and ingrained habit presupposes nothing in the way of musical ability; the fusions are merely perceived and accepted as components of the sound whose uniformities are attributable to regular and oft-recurring integrations.

[1] Cf. *78*.

It does not follow that a capacity to receive these and set them aside as commonplace would give rise to any musical impulse whatsoever. All we need to assert is that an adjustment to sounds resting upon a ready adaptation to the chief fusing intervals would be a useful basis for discriminating the more unusual components which so often are of vital significance to primitive man.

The objection that has been made against this theory (*39*) on the ground that among primitive peoples, and in the music of barbaric races, harmonic intervals, excepting the octave, are rare, misses the point of the argument; for, although some primitive peoples have not followed the course we have taken in devoloping our music, this does not show them to be lacking in a dispositional readiness to accept intervals fusing in the manner noted. Harmonics are used in the choral singing of all peoples, and the employment of the harmonic principle is evident in the frequency with which primitive duplications of melody with different voices are built upon the fusing intervals of the octave, the fifth, and the fourth. At the same time, other and dissonant intervals separating the voices are likewise found in primitive music; such, for instance, as one approximating the major second (8:9). This, of course, is not selected by virtue of its fusion. Yet its employment can be accounted for as the adoption of a bitonal chord, the interval of which had already become habitual in melodic sequences. The principle of proportional division, to which we shall return, explains the division of the octave, for purposes of melodic development,

into small, equal steps, and these intervals may be carried
over into polyphonic usage without any regard to fusion.
Harmonic effects do not become dominant until harmony is
a prime consideration of musical structure, and, as we shall
see when we come to study the nature of music, harmonic
effects are quite differently ordered from those of simple
melody; for whereas the influence of fusion is dominant in
harmony, equal intervals and their multiples seem to be the
more important constructive features in the development of
pure melody.

As for the fundamental fusions, they may well be operative in audition without influencing musical practice in any
degree.   Our conclusion, then, is that fusion depends
upon a dispositional readiness to perceive as a uniform impression any simultaneous combination of tones embracing
intervals that have been frequently encountered as inherited
and individual adaptations.   These normal combinations
are grouped and set apart from all other constituents that
may characterize the sounds as individually peculiar, or
significant.   Like the vocal formants, fused intervals tend to
occupy the background of our auditory consciousness to
which attention is less readily given than to the fortuitous
components of sound with which they contrast.   The more
uniform and regular a phenomenal structure of consciousness, the more readily is it accepted without scrutiny.   Attention being normally reserved for the unique and the
peculiar, a high level of intelligence is requisite before the
details of orderly arrangement which determine the nuances

and shades of these original integrations of attributes can themselves be apprehended and discriminated.   Accordingly, music like any other design or pattern must await an attitude of leisurely contemplation before its possibilities can be developed as an Art.   Yet from the outset the fundamental configurations of fusion serve to constitute a basic structure upon which as a ground every peculiar and unusual component of sound tends to emerge into attentional clearness.   The more fundamental fusions, the octave, the fifth, and the fourth, probably arise from inherited tendencies, while the fusions of lesser degree depend more largely upon acquired musical usage.

The experiments of Moore have shown that one may learn to accept, as consonant, intervals that do not readily fuse, and that the best of fusions, the octave and the fifth, are by no means the most consonant, in the sense of being the most musically attractive.   Other investigators have likewise shown that the order of merit among combined tones varies with the attitude from which the combination is judged.   It is appropriate to suggest, therefore, that tonal fusion is an inherent dispositional readiness to integrate tones, the intervals of which are both conspicuous and frequent in the "harmonic chord of nature"—that is, in the partial tone series. Whether or not use is made of this harmonic principle in musical practice is another matter; yet in any event fusion has a "survival value," since a tendency to blend tones of intervals which occur frequently frees the mind to analyze the more unusual and less frequent components that characterize and distinguish important sounds of the natural environment.

In advocating the harmonic theory we are not obliged to reject the data upon which Helmholtz and Krüger base their views. In so far as these data are perceived they may aid in the finer discrimination of fusing and non-fusing intervals. Nor do we dismiss the hypotheses of Lipps, Stumpf, and Watt as wholly inconsequential; for the rhythmical impulse of Lipps may play some part in the impression, the "specific synergy" of Stumpf may be an appropriate term for the integration we have described, and something like Watt's "balance" and "symmetry" does belong to the phenomenon of fusion. What we do reject is the view that fusion rests entirely upon immanent features such as "rhythms," "relations," and "balances," or upon the presence of adventitious aids such as overtones and difference-tones. Instead, we regard the integration of the harmonics as biologically conditioned, and hence selected as a neutral configuration of sound, with reference to which variations of timbre—tonal, vocal, and noisy—may more readily be apprehended and discriminated. Common factors thus come to be grouped and set off as a single fusion over against the non-fusing components of a complex sound. The integration of two tones into a fused unit is but a step beyond the primary integration of the pure tone, the vocal formant, and the simple noise. Upon analysis it reveals at least two pitches, instead of one only—as is found in the tone, the formant, and the noise. Thus fusion involves a certain com-

plexity which the simpler sound lacks. Yet this distinction
is chiefly logical; for, as Stumpf maintains, fusion persists
despite analysis. The fusion does not fall apart into two
components when we direct our attention successively upon
its two pitches, any more than the intermediate color, orange, becomes now a red and now a yellow as we shift our
attention from one to the other of these two fundamentals
of color. A fusion is a compulsory unity, and must be accepted as such. The analysis of a fusion takes place only
under the conditions of a special attitude in which attention
clarifies a single attribute in an ideally constructed setting.
The setting is furnished by the attitude which makes the
abstraction of a single attribute possible. But the same
procedure is necessary if we would analyze a simple tone.
And hence it does not appear that the perceived tone, taken
as a complete pattern, is any more "simple" than a perceived
fusion. Simplicity and complexity are, indeed, relative, and
presuppose analytic comparisons made with the aid of appropriate attitudes of observation. This applies also to the
larger groups of auditory perception, such as clangs, complete vocalizations, and the noises of everyday life; for, in
all of these, analysis will reveal many primary integrations
that can be isolated as distinct though subtle components.
So with regard to the "degrees" of fusion: these are altogether relative. We say that the octave fuses more closely
than the fifth, and the fifth more closely than the fourth.
The thirds and sixths are more equivocal, more neutral, while
the sevenths, seconds, tritones, and semitones are all disso-

nant. But these judgments are *musical*. They refer to a musical setting. They have, to be sure, their compulsory basis, for we are less adapted to the dissonant than to the consonant intervals. I have sought the reasons for this in the prevalence of these intervals among the partial tones—the harmonic chord of nature—in the order of their frequency. Thus, when we undertake to analyze and compare chords, we find that some are more readily accepted as unitary structures than others. The order, or degree, of their fusion is most obvious in the case of the fusions that occur most frequently and most strikingly in the harmonic series. The order becomes ambiguous, however, as soon as we pass beyond the fourths, because the non-musical person finds these fusions less unitary, while the musical person begins to read into what he hears all manner of musical interpretations. But with practice, and a controlled attitude, even relatively dissonant intervals can apparently be made to achieve a higher level of integration than they originally possessed. Here we have an indication of the manner in which nature by a slower course has selected the simple tone, the vocal formant, the simple noise, and the fusing interval as integrated units out of the mass of phenomenal sound with which the mind of man originally was confronted.

The harmonic theory regards fusion as an integration of tones in intervals which in the course of racial experience have occurred together so often that the acts of grouping them, and then setting them aside from the more variable components of the sound-mass, have furnished an important basis for auditory discrimination. We find here neither a new "element" nor a new attribute, but only a perceptual pattern based upon recurrent uniformities of impression and response. Like the innate uniformity of parallel finger-movements, as compared with the independent use of a single finger, which is only gradually acquired, so the timbre of a clang first affects us as a whole. The selective response to its individual components comes later; but when it does come, it is made most readily with respect to the non-harmonic components from which the harmonics are set off as a sort of background upon which unusual sounds emerge in the field of attention. The harmonics blend or fuse, because the occasion for their discernment has been less frequent and less insistent, and they retain this unity of structure even when an analysis of their components is made.

CHAPTER VII

THE CHARACTER OF SOUNDS

THE PRINCIPLES OF TONALITY

In our preceding study of tonal fusion we have discussed the relation of one tone to another in succession, and also the character a tone acquires by virtue of this relationship. We may now proceed to a consideration of tonal character, or tonality, and likewise of the character of sounds other than tones.

In a sequence of tones, the successive ratios of vibrational frequency are not a matter of indifference. On the contrary, there are principles in accordance with which sequences of tones associate themselves in an orderly progression or movement. Such sequences we call *melodies,* and their underlying principles make musical thought and musical composition possible. We must therefore try to seek out these principles before we can hope to understand the essential nature of music. But let us first note that tonality as well as fusion rests upon the sense of interval; and interval, as we have seen, is limited by our ability to discriminate a difference of volume. In a succession of tones we can not employ any interval which is not readily distinguishable in this way. If we recall that, within the musical range of tones, the liminal difference for volume is measured by a fractional increment approximating three one-hundredths of the vibrational frequency, it appears that if an octave be divided into twenty-four quarter-tone steps, the adjacent tones will be hardly discriminable as different intervals. Since the fractional increment for the semitone is about six one-hundredths, the quarter-tone interval, being half that, or three one-hundredths, is already within the liminal range of uncertainty.

There are several principles governing the alliance of tones in a sequence. The first is the *law of the tonic,* which has reference to certain reciprocal relationships among tones. The second is the *law of cadence,* or the *falling inflection,* which concerns the rise and fall of both pitch-brightness and intensity. The third is the *law of return,* which involves a demand for return to some outstanding tone in the sequence, preferably the first. The fourth may be described as the *law of equal intervals,* or the proportional division of the scale which governs the judgment of equal intervals when tones of different pitch evince like ratios of vibrational frequency. We shall now examine these principles of tonality, and attempt to formulate the appropriate law of each.

THE LAW OF THE TONIC

The phenomenon of the tonic can be demonstrated by the following experiment. Select two tones in the ratio of 2:3, as those of c and g on the piano. If we strike these keys repeatedly in succession without unduly emphasizing either, and ask ourselves on which tone the sequence will most satisfactorily end, the answer is obvious—the more satisfying terminal is c. If we repeat the experiment, selecting c and f as our two keys, it will be found that not c, but f, gives the more satisfactory ending. In each instance our judgment is determined by the law of the tonic; for c is a tonic to g, while f is a tonic to c. But what is it that occasions this apparent shift in the trend of these tones? If we examine the two ratios, 2:3, and 3:4, we find it is the ratio-number 2, or its power 4, that in each case represents the more satisfying final. Experiments with other intervals involving 2 or a power of 2 show that, other things being equal, the trend of a two-tone sequence is always towards the one whose ratio-number is 2, 4, 8, 16, 32, etc., whenever such a tone is present.

If we ask why this should be so, we may point for our answer to the dominating position of the *fundamental* in the harmonic series of the partials. The ground-tone is the tonic of all its partials. Since this tone finds itself repeated by octaves in the second, fourth, eighth, and sixteenth partials, it can exercise a dominating influence over the entire clang. This seems to be the chief reason for the tonic effect. In the case of a ratio such as 3:5, the major sixth, there is no such dominance of one tone over the other; and so far as this particular feature of alliance is concerned, the sequence is *atonic,* or indeterminate in trend.

We have here a principle of great importance in melodic structure. A return to the tonic is welcomed. By observing this principle a melody achieves both form and completeness. The law of the tonic may be stated as follows: In any sequence of tones in which one tone appears whose ratio-number is 2 or a power of 2, all the others evince a melodic trend towards it. It is welcomed when it occurs, and it serves as a satisfactory terminal to the melody.[1] Or, as Meyer expresses it: "When one of two related tones is a pure power of two, we wish to have this tone at the end of our succession of related tones, our melody" (*65*, 9).

As a corollary to this law, it also happens that we can be satisfied with an ending upon any one of the tones in a tonic chord. But this variety of ending is really much more erudite, for it presupposes an attitude familiar with the tonic, and also with its more usual harmonic accompaniments. The tonic chord being 4:5:6, an ending on tone 5 or 6 satisfies us in so far as it suggests the dominant ground-tone of the chord, 4.

THE LAW OF CADENCE, OR THE FALLING INFLECTION

But melodic trend is not governed by the tonic alone; for a melody may be a succession of tones in which no tonic appears; or it may be a succession of tones in which none of the intervals conspicuous in the harmonic series of the partial tones is present. In the former case, intervals such as 3:5, 5:6, and 9:10, have no tonic; yet if we use them, a melodic trend may still be noted. Perhaps the most im-

[1] Cf. *15*.

portant law governing such instances is the *law of cadence*. Other things being equal, finality attaches to the lower tone, an effect which is also enhanced by the decreased intensity that usually attaches to low tones. We observe this principle at work in the tendency to lower the voice at the end of a sentence, and it is likewise familiar to us in many expressions such as *zig-zag, tick-tock, knick-knack, flip-flop*, etc., all of which would sound strange if the syllables were reversed. One may also note that in each of these four illustrative instances the vowel-sound of the second syllable is that of a lower region of pitch than the vowel-sound of the first syllable of the pair. The reason for the falling inflection is probably to be found in a relief from tension, for a greater relaxation is usually associated with sounds of low pitch, and they are also intrinsically less intensive than the sounds of higher pitch.

THE LAW OF RETURN

A further principle may be formulated as the *law of return*, which depends for its effectiveness upon a more or less definite memory. The attitudinal adjustment attending the reception of the first tone in a melodic sequence is, like all first impressions, a thing of lasting influence. The law of return may be thus simply stated: Other things being equal, it is better to return to any starting-point than not to return (*15*, 33 ff.). The starting-point, however, is not in all instances the very first tone to be sounded; sometimes a few tones lead up to it. But in such cases the tone to which a return is made is likely also to be the tonic; if not, it must be otherwise emphasized by an intensive accent, or by an appropriate rhythmical setting.

A good many different factors are involved in the accentuation of tones in a melodic sequence. Both the comparative effectiveness of different intervals, and the trend of their sequence, may be influenced by the pitch-brightness, or the volume, of individual tones, or by the distance of the tones and the musical intervals employed. Thus Sterzinger (*112*) found that whenever two or three tones of equal intensity were sounded at equal intervals of time, an inherent accent attached to the highest tone, and introduced a subjective rhythm. The time of the interval was also found to be of importance, for if the two tones were identical, an interval shorter than one-half a second tended to be iambic, while longer intervals appeared to be either trochaic or spondaic.

Musical intervals were also found to alter their character according as attention was given to the volumic basis of the interval, or to the pitch-distance traversed (*113*). Thus a small interval tends to be apprehended as a *step*, indicated by a small but apparent volumic difference, whereas a larger interval is more like a *jump*, in which the tones stand clearly apart, because of the notable distance between them. The effects of consonant relationship and rhythmical accentuation combine with these differences to determine the course of the melody with its recurring intervals, including the one with which the melody finds an end.

THE LAW OF EQUAL INTERVALS

The fourth principle of tonality has nothing to do wit
trends suggestive of finality, and it operates in sequences o
tones that do not comply with the conditions of a harmoni
series. The relations we have now in mind are founded upo
equal ratios and their multiples, the unit of experience bein
a distinct interval, the ready apprehension and employmer
of which suggests a unique capacity for estimating volumi
proportions, whether they be "musical," in the sense tha
they belong to the harmonic series, or not. For instance
the Javanese employ a standard interval whose ratio i
519:596. This approximates, but is somewhat smaller thar
the simpler ratio of 7:8. Javanese melodies are built upo
successions of this interval and its multiples. The interva
is obtained by dividing the octave into five equally-tempere
divisions, each two adjacent intervals having the same rati
of vibrational frequency.

We here meet with a new principle of tonality—the or
ganization of tones upon a unit whose ratio is not include
in the harmonic series, but is one which nevertheless serve
as a recognizable interval. We shall consider this principl
at greater length when we come to the subject of the musica
scale. For the present it will be enough to note that the
employment of such intervals follows from the rathe
simple expedient of dividing an octave into a small numbe
of parts whose ratios are all alike. The "sense of propor
tion" required to establish this scale implies an ability to

recognize equal intervals and their multiples, and therefore differs from the "proportionalities" of the harmonic ratios. The employment of equal intervals obviously precludes the use of a tonic, since an octave can not be so divided and still preserve the harmonic relations.   However, it is possible to approximate harmony with equal intervals, as has been done by selecting tones in the chromatic scale of equal temperament.

Tonality is a characteristic of tones arising from a relationship with other tones, either present or implied.   Its various aspects are more or less pronounced, in accordance with the particular system of tones employed.   In this regard the traditions of European and Oriental music differ. Our common tradition emphasizes the tonic, the law of return, and the law of cadence, whereas in Oriental music the law of equal intervals supplants the law of the tonic.   Yet in both types of music tonality means that sequences of tones arouse attitudes of expectancy.   In our music, we are expectant of tones to follow that may fulfil the varied desires for a tonic, for a return to some particular tone, or for a falling inflection.   That is to say, a tone heard in a musical setting demands other tones, higher or lower, that will increase the state of tension to a certain point, and then provide relaxation by a return to some accentuated tone, or to a tonic, or to a tone of a lower pitch.   In Oriental music expectancy is based upon a unit-interval, which may be repeated either simply or in multiple form upon different tones of the scale.   Cadence and return are also involved, but the unit-interval serves as a measure for the melodic sequence. The elaboration of a melodic pattern follows the same general principles whether we employ tonic trends at the expense of equal intervals, or equal intervals at the expense of tonic effects, or, indeed, a combination of both. For just as word leads to word in linguistic communication, so tone leads to tone in musical expression; principles of orderly sequence underlie both. The word has no meaning apart from its context; neither has the tone. As the suggested or intended context gives meaningful character to the word, so the suggested or intended musical context lends character to every single tone employed.

TONALITY AS AN ATTRIBUTE

What we have here described in its specific bearing upon tones in sequences is often taken to be an elemental aspect, or attribute, of sound. Such an inference has been made by Révész (96), on the basis of some interesting experimental results which he obtained in co-operation with Paul von Liebermann.[1] As these experimental data have a direct bearing upon the question of tonality, we shall at this point review them. Von Liebermann, a person of musical training and appreciation, suffered an auditory disturbance which brought with it certain peculiarities of hearing. Révész, testing his friend experimentally, found that, in certain regions of the scale, von Liebermann confused such intervals

[1] Cf. also Meyer (66, 68).

as those of the fifth and the octave. Yet his ability to distinguish these same intervals as regards their *distance,* or span, remained intact. Révész has described the anomaly in terms of *pitch* and *quality.* According to his report, von Liebermann's sense of pitch remained normal, while his sense of quality suffered derangement. The demonstration is apparently faultless, and the question therefore arises: How can we reject tonality as an attribute, and still be able to classify von Liebermann's anomaly of hearing?

The history of the case clearly indicates that the disturbance was one of a functional order. After he first came under observation, the patient's ability to discriminate tones frequently altered. Yet the defects were always of a musical kind, in whatever region of the scale they might chance to occur. Consequently, the confusion of intervals such as the fifth and the octave may be regarded as a significant support for our theory that fusion is based upon definite integrations dependent upon certain physiological conditions. Since the fusion of tones was altered without a corresponding loss of hearing for the tones themselves, we may conclude that continued accuracy in discriminating both height in the tonal scale, and distance of interval, gives evidence that the attribute of pitch was not involved. What happened in the affected regions of the scale was the reduction of von Liebermann's musical ability to that of an unmusical person. Whenever a musical integration involving harmonics was called for, the patient was unequal to the task. He could discriminate both simultaneous and successive intervals as distances, but not as the fused or related intervals of music. His harmonic habits of musical thought were weakened, or destroyed, just as are verbal habits in certain aphasic conditions. Thus the loss of tonality in this case seems to point not to an attributive, but rather to an integrative defect.

That is not the whole story, however, for the regional nature of the defects indicated functional disturbances that were more or less definitely localized in the tonal series. Since the phenomena were binaural, it would appear that the disturbance must have been central. The difference in integration taking place at certain points of the scale under similar objective conditions, and the apparent identity of units of distance in the normal and in the paracusic regions, suggests that the patient's disability exclusively concerned the construction of musical intervals, both simultaneous and successive. Tests indicated that his judgments of vocality and of tonal distance were both normal, the only defect being that of tonality or musical integration. Had the paracusis been monaural instead of binaural, we might suppose that it was localized in the mechanism of the inner ear, which of itself would give no support to a functional theory of the musical interval. Since, however, while binaural agreement persisted, the affected regions varied in the course of the disease, we shall not be far wrong in concluding that the derangement was functional, and that it concerned the central nervous system.

What Révész calls "quality," therefore, need not be accepted as an attribute, but rather as the result of a perceptual integration in which pitch-brightness and volume were primarily concerned. We have evidence in von Liebermann's case of tones that can be heard, although they neither fuse nor establish musical intervals. Judgments of distance, height, and vocality were normal, while those of fusion and tonal relationship were abnormal.

According to our hypothesis this implies no sensory defect, but a central disturbance of function with respect to certain integrations among the phenomenal attributes of sound. Instead of justifying the inclusion of tonality among the attributes, von Liebermann's case supports our theory of the musical interval as a phenomenal structure based upon a certain integration of pitch and volume, the physiological conditions of which have their origin in the brain.

ABSOLUTE PITCH

Another phenomenon bearing upon the question of tonality is that of *absolute pitch*. A few exceptional persons, probably all of whom have had a musical training, enjoy the gift of naming a tone correctly when it is heard. This ability appears to be most accurate in naming the place of the tone in an octave-sequence. It is somewhat less accurate in naming the proper octave. The sense of absolute pitch is not a judgment in which the observer compares the tone heard with a standard he has in mind, and which he can imagine or imitate vocally; for judgments of this sort are erudite and rational as compared with the immediacy of absolute pitch. Furthermore, the accuracy of absolute pitch is always impaired whenever the hearer tries to reason out his judgment.

Cases of this remarkable ability have been carefully studied, leaving no doubt as to the genuineness of the phenomenon. Must we not, then, take absolute pitch to be an evidence of inherent tonality? These persons seem to be able to detect c's as c's, d's as d's, etc., as though the original positions of these tones in the musical scale were immediately apprehended. Yet the judgment must involve something more than an intrinsic quality, otherwise these outstanding tones would be absolute with reference to their vibrational frequencies, whereas they are absolute only with reference to certain intervals. When we consider, too, the non-harmonic intervals employed in Oriental music, it becomes even more improbable that absolute pitch has anything to do with intrinsic tonal quality. The phenomenon, therefore, appears only with reference to a scale with which the individual who possesses this talent is familiar. As such, absolute pitch may be as common to Oriental musicians who use non-diatonic scales, as it is among our own musicians; the records of Malu music obtained by C. S. Myers (76) give evidence of this among a very primitive people. In every instance of the possession of absolute pitch an interval is implied; although the phenomenal character of the tone-structure is a distinct and important criterion of the judgment. Köhler (50) found that he was able to improve his ability in naming the place of tones in the scale when he gave special heed to timbre and "tone-body"—meaning by the latter term, we should say, an integration of volume and brightness.

Tests of accuracy in the judgment of absolute pitch give support to Köhler's emphasis upon timbre; for J. W. Baird (5) found that judgments based upon pure tones, and upon tones of the human voice, were much less trustworthy than those upon tones of several common musical instruments. Among all instruments, the tones of the piano were shown to be the most accurately judged. Pure tones of course lack timbre, while the human voice, though very rich in partials, varies from individual to individual. Absolute pitch is found to be most accurate in judging tones whose timbre is both striking and uniform. Baird also found that while the confusion of octaves was more frequent than that of any other intervals, fifths and fourths were less often misjudged than thirds and sixths. Doubtless we see here the influence of what Köhler has called "tone-body." The tone must find its place within the setting of an octave, and thus establish its interval before its pitch can be named. This place is marked by its volume and its brightness rather than by its musical effectiveness. Harmonic relations play a less important part than position, because they more largely depend upon musical thought, and therefore must be less immediate in their effects. Now the fourth and the fifth are both harmonic intervals, but are also more prominent divisions of the octave than are the thirds and the sixths. As a measure of the octave they assume, therefore, a greater significance, which explains why they are less often misjudged than are the thirds and sixths. In his records of some very primitive vocal music among the Murray Islanders, Myers has noted that, although the sequences of the tones employed were non-harmonic, there was a tendency to approximate intervals of the fourth and fifth, but never those of the thirds and sixths. He also reports memory for absolute pitch among these savage vocalists in their capacity to return, after singing several non-harmonic intervals, to the tone from which they started. We may infer from this that reference to a unit-interval is of greater importance in judgments of absolute pitch than any reference to harmonic character.

The phenomenon of absolute pitch is thus explained in a general way by inferring that certain individuals have been led to regard the attributes of volume and brightness in tones of a certain timbre, and have come to associate these with the scale of a familiar instrument. In this way they are able to recognize the tone immediately in its proper place within the octave, and they may likewise be able to locate the particular octave in which the tone occurs. That different timbres should affect our judgment is explained when we consider that a change of timbre always alters the volumic pattern and total effectiveness of a sound. Yet it is as easy to become habituated to one more or less constant timbre as it is to another. The important thing is that the timbre shall not vary too greatly, and that there shall be "richness" enough in the tones judged to render the pattern of the clang impressive. A refined sensitivity to different clang-patterns seems, therefore, to be the basis upon which judgments of absolute pitch are made.

THE BASIS OF TONALITY

Phenomena such as have been described tend to show us that tonality is not an attribute of sound, but a perceptual characteristic of tones in their sequential setting. Upon which one of the attributes tonality chiefly rests it is hard to say; all are in some measure involved, though pitch, being the most striking feature of tonal order, is often used synonymously with tonality. The absence of a dominant pitch distinguishes all sounds that are not tonal in character, and because of this absence we can form no melodies with them. But volume and brightness also play important parts in tonality. The sense of equal intervals is primarily determined by volumic differences, since the discrimination of volumes depends upon a proportional increment. It is noteworthy that a considerable range of high tones can be distinguished by pitch-brightness even though these tones have no musical significance. As we pass upwards in the scale, tones become less and less voluminous, and the discrimination of their respective volumes becomes more and more difficult. Among high tones, discriminations of pitch and volume both suffer, the pitch tending to appear lower than it should. With low tones, on the contrary, the pitch tends to be sharped. There is also this further difference between the extremes of the scale: among high tones, the pitch-salient is more prominent than the volume, whereas in the lower portion of the scale, volume seems to suffuse and blur the pitch-salients. For this reason both very high and very low sounds are more like noises than they are like tones; and we can establish neither equal nor harmonic relations between them.

If simple proportionality and the harmonic relationships are the chief features of tonality, both volume and pitch must be involved. But even if these attributes are more important than the others, intensity, duration, and brightness are by no means negligible. We have referred to the principles of *return* and *cadence* as likewise contributing to tonality. Duration, dullness, volumic size, and decreased intensity are all associated in cadence with the feeling of relaxation which appears to occasion this tendency; while the law of return operates with volume, intensity, duration, and brightness, all of which assist in emphasizing the tone to which a return is afterwards sought. We are therefore safest in concluding that tonality rests upon all the attributes of sound.

THE CHARACTER OF VOCABLES

Has the conception of tonality a meaning only for tones? Are there not analogous effects among the vocables and the noises? While the pitch-attribute of the formant is not a pronounced salient, it is evident as a regional effect; and the regional relationships of the vowels *u, o, a, e, i* follow an ascending pitch-order. In this upward progression of the vocalic sounds, both pitch and volume are implied; although, the pitch being regional, the vocalic sounds can not be subjected to the finer discriminations that lead to the precision of harmonic and equal intervals.

Further investigation is necessary before we can say more about the character of the vocables. Habituation to a subdued or non-salient resonance in certain regions of the scale, when associated with an appropriate fundamental or "voice-tone," is apparently characteristic of the vocalic sound, but its precise pattern in terms of attributive integration, and the orderly sequences of vocalizations, is a problem of phonetic research into which we can not here enter.

THE CHARACTER OF NOISE

In the realm of noise we also find effects analogous to those of tonality. While the regional relationships of various noises have not been adequately investigated, the structure of the consonants, as defined by Stumpf's analyses, suggests for each a definite regional resonance analogous to that of the more tone-like vowels. Because of the physical complexity of its stimulus, noise has always seemed a very difficult subject of analysis. But the recent investigations of Abraham (2, 3) on simple noise, with his demonstration of the independent variability of wave-length over against vibrational frequency, have opened a way to further experiment in which the physical conditions can now be better understood and more accurately controlled. While the pitch-salients of all noises are confused and irregular, the volumic pattern and the aspect of brightness attributable to wave-length afford more constant factors. If the volume of the lowest component embraces all the other constituents of the sound, a consideration of different noises with regard to their volumic sizes is a possible means of analysis. Relationships of this order are already indicated by the sequence of the consonants, and may be extended to include all other general types of noise. These types must all be distinguishable within the range of volume-differences, their special uniformities being, of course, other than those of the octave-relationship with its dependent orders of harmony and equal interval. Aside from volume, the attributes of intensity, duration, and brightness are likewise conspicuous in the description of noises. In the absence of dominant pitch and volumic uniformity, one might expect these other attributes, especially that of brightness, to be the significant features by which noise is characterized.

SUMMARY

The classification of sounds into tones, vocables, and noises is not an elemental, but a perceptual distinction, dependent upon integrations of the attributes of sound. The character of a sound, like the character of a man, is a matter of *conduct*, of the way it behaves or functions; and this, in turn, is not merely a matter of the attributes it possesses, but rather of the way in which they integrate into perceptual patterns. Each of these three types of sound possesses the same attributes: pitch, volume, intensity, duration, and brightness; yet in their varied integrations there arise distinct characters of sound. Nor can it be said that certain attributes are more important than others, although we have studied these integrations chiefly with reference to pitch and volume. The co-operation of intensity, duration, and brightness, is equally important in the total integration, for each is a discriminable aspect, with an inherent variability which contributes to the character of the sound as indicated by its influence upon other sounds. We can not enter into the details of this manifold of integrative patterns, but may note that herein lies the explanation of music, of our system of vocal sounds, and of the as yet uncharted region of significant noise.

The character of any sound is inherent in the pattern of its attributes, the foundations of this character being determined within the sound itself. Possessing such and such attributes of pitch, volume, intensity, etc., the sound is perceived as a certain kind of experience. Not that these attributes integrate themselves without reference to the perceiving mind, but that the mind must accept the sound pretty much as it is. The affinities of the attributes, being inherent, are implicit rather than explicit; that is to say, they require no analysis to demonstrate their existence. Prior to all cognitive analysis these percepts arise through "practical judgments," responsive activities in which a certain impression of sound is accepted as a unit. At this level of response, the precise nature of the impression is not analyzed, nor have any of its terms been raised to the status of "free" ideas. As Stumpf remarks, fusion exists as an immanent relationship prior to any attempt at analysis. When we do analyze the chord, our attention is drawn to the idea of fusion that is inherent in it, and yet this impression of fusion continues to exist even when the analysis is achieved. Tones, vocalic sounds, and noises exist perceptually, because the attributes which determine their essential components are implicitly integrated in a manner that lends to each particular sound its distinctive character. When explicit analysis is undertaken, one discovers for the first time how these attributes enter into correlation with one another, and with their contexts; then for the first time we begin to understand the elements of music, of speech, and of significant noise.

By processes of abstraction we are able to concentrate upon fusion, tonality, vocality, or noisiness. In so doing we acquire ideas concerning these varied characters, and an ideational setting for their occurrence and behavior. This enables us to exercise a certain constraint over them, so that the same sound may serve at different times as a tone, as a vocable, or as a noise, according to its context. Relationships which were at first implicit now become explicit; they are no longer tied to the bare perceptual existence of the sensory data, but are re-created in the act of selecting and rejecting whatever constituents of the sound the mind may choose to use or to discard. At this level of consciousness, the influence of personal design is felt; being engaged in musical thought we can detect tones consonant with the pattern of our musical progressions in even so refractory and unmusical an instrument as the kettledrum, because the attainment of a musical attitude is conducive to the apprehension of musical elements in whatever sound is heard. Yet in this exercise the mind is never quite free; it must defer to objective conditions so long as it is primarily engaged in perceiving sound; but for the implicit relationships of the attributes as they exist in the sounds we hear, it would be impossible to reach the heights of theoretical judgment to which the creations of music and language belong.

CHAPTER VIII

THE MUSICAL SCALE

THE ORIGIN OF THE SCALE

A MUSICAL scale is an ordered series of tones derived from the tones used in the composition of a melody. One commonly includes in a scale only those tones which characterize a certain *modus*, or musical effect—although the melody may introduce extra tones, known as "grace-notes," or accidentals, which do not properly belong to the scale. Furthermore, a scale obtained by selecting and arranging in orderly progression all the different tones that are prominent in a given melody may embrace intervals neither characteristic of nor suitable to the modus in question, and such a scale might also lack certain intervals which are used in other compositions of the same modus. Thus it appears that the scale registers in an orderly sequence, not only the tones actually employed in a certain type of composition, but also whatever other tones may be needed to complete the modus in which the melody was composed; for melody is not an aggregate of tones, but the moving pattern of a succession of intervals. The intervals, then, which characterize the peculiar modus of the melody are of even greater moment than the tones themselves, if these be regarded as elements; for the same melody can be reproduced by transposition with quite different tones, provided only that the same intervals are retained.

Since the principles determining the selection of musical intervals are relatively few and simple—as has already been made plain—the number of possible scales of a musical order is not without definite limits. We have already shown that all musical expression tends to fall under one of two fundamental types: (a) the employment of equal intervals and their multiples; and (b) the employment of harmonics and their derivatives. It is, however, possible to produce a scale which answers with sufficient accuracy to the demands of both principles—for instance, our own chromatic scale of equal temperament. Yet such a combination to meet diverse demands is usually but a practical expedient of the instrument-maker. While it is more than a happy accident that the chromatic scale so nearly conforms to the principle of equal interval that in tuning our pianos we can procure a close approximation of the diatonic scales with every key, the musician knows the difference between harmonic and equal intervals, and makes imaginative allowances for the discrepancies he hears. With an instrument self-tuned like the voice or, to some extent, the violin, the artist will, of course, make his intervals what he desires them to be—harmonic or equal, according to his needs. Thus no discrepancies are felt, and both harmonic and equal intervals may be woven into a single complex pattern, provided that the musician is competent to apprehend and make proper use of each of these principles as he employs it. If he lacks ability to discern the difference between harmonic and equal intervals, he soon gets off the key, and wanders into a strange modus which destroys his musical pattern.

While the principles of both harmony and equal interval are employed in all music, a consistent use of one is inimical to an equal consistency in the use of the other; for there can be no single musical scale embracing all the possible intervals of a musical composition, in which every tone is related musically to every other tone. In the construction of a standard scale the selection must be limited to those tones within an octave which bear such mutual interrelationship as will provide the largest number of melodic sequences consistent with a certain musical attitude, or modus of expression. In this manner the scales of different peoples have arisen and become standardized—such as the Greek scale, the mediæval modes, the major and the minor scale, and likewise a large variety of scales met with in the Orient. Yet, as we shall see, all these diverse scales trace their origin to different applications of the two principles of tonality which we have termed the principle of equal interval and the principle of harmonic relationship.

Let us first trace the origin of our own diatonic scales, all of which are derived from the employment of intervals that show the effects of fusion. In other words, the intervals of our scales are selected in accordance with the harmonic principle which has regard for tones that sound well together. Yet, despite this fact, it would be incorrect to deem the origins of our music harmonic; for, in the light of historical evidence, it appears that all music originates in simple melody, that is, in sequences of related tones without regard for a harmonic accompaniment. Nevertheless, the relationships that have dominated in the melodic evolution of Occidental music are the ones which we have called harmonic, and the originators of our music seem to have builded better than they knew; for when they chose to divide the octave into these particular intervals, they opened the way, not only for melody, but also for a future development of harmony.

Eastern peoples, on the other hand, who chose to adhere more closely to the principle of proportion in the selection of their intervals, renounced this prospect, and, instead, laid the foundations for a melodic evolution with which harmonic accompaniment and polyphonic enrichment could never become associated. In spite of these fundamental differences, the natural setting of all scales is the octave, and this interval, at least, seems to be common to both types of musical evolution; for we find it in the scales of the Orient as well as in those of the Occident. The identical tonality which attaches to tones in octave-relationship is, perhaps, sufficiently obvious to be inescapable in any musical development. Thus, although the tonic effect may be neglected, the octave still remains a natural framework embracing all the tones determined upon and used in any musical sequence whatsoever.

In very primitive music, however, even this framework may be lacking; for melodies have been found consisting of simple themes, based upon intervals of approximate equality, which do not span an octave. To these equal intervals there may be added a few harmonic intervals, such as the fourth and fifth, and even the octave; but the two types of interval stand apart because the harmonic intervals do not necessarily embrace the equal intervals employed. As music develops, however, a reconciliation between harmonic and equal intervals is requisite. Then it is that one or other principle becomes dominant. The octave emerges in choral singing, when the male voices take a lower, and the female voices a higher tone. If the fourths and fifths are likewise preserved, the direction towards a diatonic mode is indicated, whereas if equal interval is the dominant influence, the harmonic effects are neglected. The songs of the Malus, as reported by Myers (76), indicate this primitive stage of development before a reconciliation has been effected. The melodic themes of this music are built upon the principle of equal intervals, yet harmonic intervals are also employed whenever a larger step is taken.

THE GREEK SCALE

Our own musical scale traces its origin to the Greeks, its characteristic feature being the use of the whole-tone and semitone intervals. The beginnings of the "diatonic" scale are shrouded in obscurity, although we have several early treatises on Greek music,[1] and also a few fragmentary records of the music itself. But even in the fourth century B.C.

[1] Cf. 120.

music was an ancient art, to the principles of which, based upon tradition and usage, an exact and scientific expression could not readily be given.

Since the time of Pythagoras (*ca.* 500 B.C.), the symbolism of numbers had exercised an important rôle in Greek philosophy, and it was not unnatural that the art of music, produced by stringed instruments and flutes, whose lengths, frets, and vents could be measured and expressed numerically, should have been a fruitful source of speculation. Indeed, the harmonies of these tonal sequences were extended to include even the harmonies of the heavenly bodies; thus, the characteristic numbers of the musical ratios achieved a far wider significance than music itself could ever have warranted. Speculation of this sort, when applied to a musical practice already elaborate in its development, has a marked tendency to make the facts fit the theory. Consequently, the music actually employed by the Greeks is not always faithfully described in the notes and treatises on the subject which have come down to us from antiquity.

One of the obstacles to our study of Greek music is the very scanty fragments of ancient musical transcriptions which are extant. According to Barry (*8,* 578) the only authentic transcriptions are the *Aidin Epitaph,* published by Ramsay in 1883, the *Ashmunen papyrus,* giving a fragment of the lost score to Euripides' *Orestes,* and two ritual hymns discovered by Homolle at Delphi in 1893. There is also evidence which suggests that the notation of Greek music was not developed until the early part of the second century,

B.C.; this would explain why so many earlier choruses have come down to us without any indication as to their musical setting. To attempt to piece out a theory from these scanty fragments, even with the aid of such theoretical references to musical art as are to be found among the classical writers, is a very puzzling task. But enough at least is known to trace back the evolution of modern music to this source, and also to gain some insight into the origin of the harmonic scale.

The Greek scale seems to have been determined in its general form by the employment of three intervals; the fourth, the fifth, and the whole tone. The first two were recognized as consonant intervals, and were determined as such; while the whole tone, a dissonance, was defined as the difference between a fourth and a fifth. The earliest Greek scale seems to have been formed by the combination of two lyres of four strings, each embracing an interval of the fourth, with a common string serving to give the highest tone of the lower and the lowest tone of the higher tetrachord. The two tetrachords thus furnished a "conjunct" scale, as compared with the "disjunct" scale of eight tones which came into use only when the two tetrachords were separated by a whole-tone interval. The common, or conjunct, tone of the early Greek scale assumed a considerable importance; it was termed the *mese*, or middle, and served as a kind of tonic for the whole scale. Thus the order of tones in the scale seems to have been a descending one from the *mese* to the lowest tone, thence upwards to the highest tone, and back again to the *mese.* We have, then, as a starting-point a scale in which three tones are fixed by the two intervals of a fourth that were employed; we may designate these in modern notation as e-a-d. The completion of these tetrachords, by the addition of two strings to each, followed in accordance with three different principles, each of which furnished a different scale: the *diatonic,* the *chromatic,* and the *enharmonic.* The first of these filled the span of each fourth with intervals of two whole tones and one semitone, just as in our modern major scale; while the other scales employed two adjacent intervals, approximating a third-tone in the chromatic, and a quarter-tone in the enharmonic, to which was added a larger interval sufficient in size to complete the span of the fourth.

Normal Diatonic   Normal Chromatic   Enharmonic

Tetrachords

The foundations of the enharmonic and chromatic scales are obscure, and their bearings upon our own scale are less direct because they employ intervals smaller than any we are accustomed to use. There is some evidence, however, to indicate that the enharmonic scale, which proceeds through two quarter-tones and a *ditone,* or double-tone interval, to achieve the fourth, was in use at an earlier time than the diatonic scale; which raises an interesting historico-psychological question. How did these small intervals come into music? At present, the interval of a quarter-tone is barely

felt to distinguish one tone from another, because it is so
near the threshold of a noticeable difference of volume,
which is essential to the definition of a musical interval.
While we can not deny that the Greeks actually employed these small intervals, it is possible that Greek usage
was less accurate in practice than the theory of the scale
would suggest.   Thus, Plato writes that "in all lyre-playing
the pitch of each note is hunted for and guessed; so that it
is mixed up with much that is uncertain, and contains little
that is steady" (*88*).   A possible explanation for this uncertainty is suggested by the close relation of Greek music
to vocalization.   Monro tells us (*72*, 119) that Greek music
was primarily vocal, instrumental music being looked upon
as essentially subordinate—an accompaniment, or at best an
imitation, of singing.   In the view of the Greeks, the words
were an integral part of the whole composition; they contained the ideas or feelings, while the music, with its variations of rhythm and pitch, furnished a natural vehicle for an
appropriate expression of these feelings.   Purely instrumental
music could not do this, because it could not convey ideas
and impressions fitted to be the objects of feelings.   Now
vocalization is not so much a matter of precise intervals,
as of gliding or *portamento* changes.   The Greek theorists
realized this, and Aristoxenus, a pupil of Aristotle, begins
his treatise on *Harmonics* by differentiating the continuous movement of the voice and the discrete movement of
tones proceeding by intervals.   We can perhaps infer from
this that the small intervals were originally introduced and

employed in order to permit of slight variations in the tone of a syllable or succession of syllables in vocalization. Yet it would be unwise to assume too hastily that the Greeks did not possess a refined sense of discrimination for small intervals. Probable though it seems that quarter-tones were originally introduced in order to provide the *portamento* glide of a syllabic utterance, it is still possible that the strings of the lyre which reproduced these intervals were tuned with an accuracy far greater than we should to-day be able to achieve by sense-perception alone; for we must remember that the Greeks were concerned only with pure melody, whose progressions emphasize the interval in a way that harmonic music does not. The use of small intervals by Oriental peoples is also a fact too well-attested to permit us to deny the possibility of their accurate perception even when they approach very closely to the threshold of discrimination for intervals. While it does not follow that these people were necessarily more sensitive than we, their musical perception, being formed on different patterns, may still have been much more refined in the matter of tuning their instruments.

As to the development of the diatonic genus, this comprised in each tetrachord a semitone followed by two whole-tone intervals; hence the "conjunct" scale embraced but five whole tones, and therefore did not attain the full span of an octave—the scale being approximately as follows: e, f, g, a, bb, c, d. It has been suggested by Curtis (*21*), upon the evidence of a passage in Aristotle's *Problems* (xix, 18), that the octave may have been supplied in this scale by playing the first harmonic or overtone of the lowest string. "Why," reads the Problem, "is the concord of the octave alone sounded [on the lyre]? For they *magadize* that, but no other." Curtis also notes that the same method of "magadizing" is used by Welsh harpists to increase the tonal range of their instruments.

By the introduction of this octave-tone we obtain the true diatonic genus. The hexachord, or six-interval scale as given above, contains two fifths, f to c and g to d; but with the addition of the octave e¹ we also have a fifth from a to e¹, and with it the succession of tones familiar to us in the diatonic scale—namely, groups of two and three whole-tone intervals separated by half-tone intervals. The scale we have used for our illustration is unlike the usual diatonic scale, however, in that it starts with a semitone; this is one of the characteristics of a Greek "mode." Other modes were obtained by starting the scale with different tones. Since this procedure involves a rearrangement in the order of the tones, there occurs among these various "modes" the one with which we are most familiar, namely, the major diatonic, whose succession of intervals is two whole tones, a half-tone, three whole tones, and a half-tone.

With the whole-tone interval defined as the difference between the harmonic intervals of the fourth and fifth, the chief rules for the construction of the diatonic genus are laid down. The semitone, the only other interval to be explained, is that increment which is needed in addition to two whole tones to constitute a fourth, and which in addition to three whole tones will constitute a fifth. Since the fourth and the fifth can be very accurately determined by the ear, the whole tones and semitones take care of themselves.

In order to understand the selection and arrangement of these intervals, we must refer to the principle of harmonic relationship, with its trend toward the tonic. Considering the harmonic series of partial tones, we find that the most conspicuous relations are those of the first four partials, whose ratios are 1:2:3:4. Being at the same time the lowest tones in the order of the partials, these four are usually the most intensive, or audible. If we regard the construction of these relations as based upon unique integrations, such as those previously described as characterizing fusion (cf. p. 140 ff.), we have an added reason for the prominence attaching to these particular intervals; for among the first sixteen partials, as already noted, the octave (or 1:2 relation) occurs eight times—1:2, 2:4, 3:6, 4:8, 5:10, 6:12, 7:14, 8:16, the fifth (or 2:3 relation) occurs four times—2:3, 4:6, 6:9, 8:12, and the fourth (or 3:4 relation) also occurs four times —3:4, 6:8, 9:12, 12:16. Assuming that the frequency of certain proportions constitutes a basis for patterns of sound which may be called *orthogenic*, it is not remarkable that the octave, the fifth, and the fourth should have exercised so commanding an influence in the formation of the Greek scale. The full complement of this seven-interval octave, with its successive groups of three and two whole tones separated by semitones, can be readily traced by what is known as the Pythagorean derivation.

With the aid of a monochord Pythagoras is said to have discovered the mathematical ratios for the tones of the scale. When the single string is divided in halves, thirds, fourths, etc., the tones resulting from these divisions are, of course, those of the harmonic series. The whole string gives the "fundamental," the half gives its octave, the third gives the fifth above, while the fourth gives the still higher interval of the fourth, at the same time completing the second octave. The Pythagorean derivation is commonly stated as a rule for tuning instruments by the employment of successive fourths and fifths.[1] Thus, from a *mese* c, symbolized by the number 2, we find a tone in the interval of a fourth below, whose ratio-number will be 3. In our scale this would be the tone g. Taking g as a new point of departure, the fifth above it would be indicated by 4½, but since, for musical purposes, octave-notes are identical, we can dispense with mixed numbers, and regard 9 as the ratio-number for this tone. Its place in the scale then corresponds to our tone d, and may be introduced as a step above the original *mese*, or c. The fourth below d will then have the ratio-number 27, and is the tone corresponding to a. From a we get e with 81 as its ratio-number, from e, b with

---

[1] A. J. Ellis points out, however (35, note, 279), that since the Greek scale was derived from a tetrachord embracing divisions of the musical fourth, tuning was probably based, historically, upon this interval, and not upon the musical fifth. As the two intervals supplement each other in the formation of an octave, the two methods of tuning would achieve the same result.

243, and from b, f with 729—in reality a sharped f. This completes the scale as shown in Fig. 17.

These ratio-numbers are employed as *symbols,* and are not to be understood as a direct expression of vibrational frequency. Each number is capable of multiplication by 2 or a power of 2 without altering its significance as a relational factor. Thus the 9 of d may be multiplied by 8 to give 72, which in relation with the 81 of e reduces to the

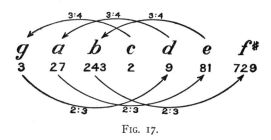

Fig. 17.

interval 8:9. Proceeding in like fashion with the other intervals, we find the Pythagorean scale to be so constructed that intervals of the fifth and fourth can be established with every tone, while the ratios of the thirds and the sixths do not appear at all. Furthermore, the scale has but two kinds of intervals between adjacent tones: one being the interval of 8:9, which obtains between c and d, d and e, e and f#, g and a, a and b, while the other approximates 19:20, and obtains between f# and g and between b and c. Accordingly, a scale based upon successive fourths and fifths provides seven tones and six intervals within an octave, and the intervals fall into two groups, one of two, and the other of three, intervals, whose ratios are uniformly 8:9, with a smaller interval approximating a semitone separating the two groups.

This is the Pythagorean explanation of the whole-tone and semitone intervals, and of the order of their employment in the diatonic scale. But it can not, of course, be supposed that the Greek scale was originally derived through any such conscious mathematical procedure. What Pythagoras demonstrated was that one of the Greek scales could be standardized as a series of tones, in which for every tone another can be found in the fifth place above it, whose ratio approximates the harmonic 2:3, and likewise one in the fourth place below it, whose ratio approximates the harmonic 3:4. As for the actual tuning of the Greek instruments, this doubtless was done by ear. The principle of equal division may then have been followed in determining the semitone intervals as approximately one-half a whole-tone step. The particular whole-tone interval that was employed has, however, a very special significance; because by chance or otherwise it happens to be a member of the harmonic series. The ear of a trained musician can readily and accurately discern it, and doubtless did so in the remote days of ancient Greece. The whole-tone interval of 8:9 also involves a tonic effect upon the note 8, since this number is a power of 2. Remembering that the Greek scale was commonly read downwards, we can perceive a special reason for the uniformity of this interval, in that the trend of each whole tone is always towards a tonic. The semitone, on the contrary, being non-harmonic, suggests its origin as a mere division of the whole tone. Indeed, the smaller intervals were probably never so definitely fixed as were the larger and more important ones; and, as already stated, they were sometimes divided again into "enharmonic," or quarter-tone steps.

In our modern scale the semitone is expressed by the ratio 15:16, a harmonic interval with a tonic trend towards the upper tone. It is not a "neutral" or indifferent interval, such as seems to have been the case with the one employed by the Greeks. In the earliest music, both harmony and equal division were probably effective in determining all intervals. But the harmonic effects, being chiefly those of the fifth and fourth, were necessarily much more restricted than are the varied harmonies embracing the thirds and sixths employed in modern music. Thus the influence of a tonic "key-note" is not conspicuous in the Greek scale, nor is it in the "modal" scales of mediæval Gregorian music, which carried on the Greek tradition. The position occupied by the tonic as a fourth from the bass made it a tone to which the melody frequently returned in the course of thematic elaboration, but the melody always ended on the dominant or lowest tone of the scale. According to Barry (*8*, 582), "this rule of cadence-structure was inviolate, and formed a criterion for the genuine"—which again indicates the close connection of melodic sequence and vocal expression, because in speech we naturally end our sentences with a falling inflection of the voice. When innovators defied the law of cadence by ending their melodies on the tonic, they were taking a step towards the liberation of music from its vocal setting, but in so doing they were decried by the classicists, who persisted in regarding music merely as an embellishment of the spoken word.

THE MODERN DIATONIC SCALE

Not until the musical effects of thirds and sixths became recognized did the norm of equal division give way to a more complicated distinction of tonic, and atonic or neutral, intervals. The increased employment of fusion by combining tones in chords, as well as the varied effects of the tonic relations in a succession of tones, gradually led to alterations in the intervals, so that these simultaneous and successive effects might be musically enhanced. The semitone step was thus transformed into the harmonic interval $15:16$, to accommodate which, two of the whole tones had to be reduced from "major-tone" intervals of $8:9$ to "minor-tone" intervals of $9:10$. This brought about the achievement of our present diatonic scale, reckoned as follows from a key-note, or tonic, upwards:

c	d	e	f	g	a	b	c^1
24	27	30	32	36	40	45	48
(3)	(27)	(15)	(2)	(9)	(5)	(45)	(3)

The numbers first given indicate a sequence of vibrational frequencies that would produce this scale, while the symbols,

or ratio-numbers, are indicated in the parentheses below. It
will be found that the successive intervals of this scale are
as follows:

c–d	d–e	e–f	f–g	g–a	a–b	b–c^1
8:9	9:10	15:16	8:9	9:10	8:9	15:16

Here arises the distinction of "major" and "minor" music.
In the Greek scale, the chief intervals within the octave were
the fifth and the fourth, supplemented by the smaller intervals of whole tone and semitone. In the modified scale,
harmonic thirds and sixths are introduced. These are the
ratios of 4:5 and 5:6 for the major and minor thirds, and
3:5 and 5:8 for the major and minor sixths. These ratios, in
which 5 always appears, are not represented in the Greek
scale. The comparative musical effects of major and minor
intervals are peculiar, in as much as these intervals possess
a similar character, except that the major is an "increased"
interval, while the minor is a "decreased" interval. Aside
from the thirds and sixths, major and minor intervals also
occur in the seconds, or major and minor tones, 8:9 and
9:10, and likewise in the sevenths, 8:15 and 9:16. The
difference between major and minor music is a difference in
the prominence given to one of these alternative sets of
intervals. The minor scale is revealed, when, instead
of taking c as the key-note, we start with a, the important
feature of this sequence being that the minor third, a-c, pre-
cedes the major third, c-e; and it is this prominence of the

decreased interval over the increased one that gives to the scale what is known as its "minor" effect.

Much has been said and written of the sadness attaching to minor music, and the corresponding gladness of the major mode; for these conventions are firmly rooted in our music. But they are only conventions. The Japanese, for instance, employ decreased intervals analogous to those of the minor mode to express joyous feelings, and mediæval composers did the same in the composition of tunes both secular and religious. In the sixteenth and seventeenth centuries the dance, as well as operatic and lyrical music, was predominantly minor, and the minor mode appears to be a conspicuous feature of folk-music, especially the popular folk-airs of Greece and Brittany (*20,* 135).

The modifications which gave rise to the modern diatonic scale made possible a greater number of tonic effects among a larger number of tones. With but three exceptions, a musical relationship of some sort is established between every two tones of this scale; the exceptions are d-f, 27:32; d-a, 27:40 and f-b, 32:45. If we regard the harmonic series as limited to the audible partial tones, these three ratios are non-harmonic. Yet they are not without musical interest for the first of them closely approximates a minor third, and the second, a fifth, while the third, f-b, has the effect of an equal interval known as the *tritone,* since it embraces three whole-tone steps. Marking as it does an almost equal division of the octave into two parts, the tritone has a special significance, even though it lacks harmony in the strict sense of that term.   Reference to the tritone is found also in Greek music, and a *Hymn to the Sun* dating from the second century A.D. contains the tritonal passage five times (*120, 73*).

While the appearance of the tritone (f-b) in the diatonic scale may be regarded as a mere exigency of scale-construction, the employment of the interval in melody does not necessarily presuppose the feeling for a more harmonic interval, such as the augmented fourth, 5:7, as Stumpf (*ibid.*) tries to make out.   If equal intervals are as original in their effectiveness as harmonic intervals, the unique effect of the tritone as half an octave may determine its selection in a melody, quite apart from any approximation to harmony. The same is true of the use that can be made of the semitone, the whole tone, and their multiples, so that whereas the intervals of the diatonic scale are chiefly harmonic, some of them can also be used as equal intervals; and our contention would be that they often are so used in passages of the semitone, the whole tone, the ditone, and the tritone. Wherever the character of the melody demands harmonic intervals, the tritone is rejected, the ditone is apprehended as a third, the whole tone as a tonic interval 8:9, and the semitone as 15:16.

Likewise in the case of the intervals d-f and d-a, although they do not possess the exact ratios of a minor third and a fifth, the hearer can readily substitute an imaginatively corrected interval for the slightly mistuned one that he hears. With many musical instruments, such as the violin, and with the human voice, the musician can control the intervals in his performance. With other instruments, where this control is impossible—as with the piano or organ—the ear of the listener will supply a subjective correction rather than detect a mistuning. Corrections of this sort are by no means difficult, as is shown by the fact that the standard scale of instruments whose vibrational frequencies are not under the control of the musician is "tempered" in such a manner that not a single interval, excepting the octave, is quite "true" to the laws of strict harmony.

THE "TEMPERED" SCALE

By "tempering" the scale, free modulation from one key, or scale, to another is made possible. Since the diatonic scale, as described above, requires a fixed order of the whole tones and semitones, evidently it would permit no duplication of its proper relations of tone with tone in any other key than the one for which it was laid down. Yet, in the very nature of this scale, a new tonic, other than the key-tone, will from time to time be announced, and whenever this occurs, the original key-tone must lose its dominating influence. In the Pythagorean scale every tone may be a tonic for some other tone in the relations of fourth and fifth, except that a true fourth is not provided below f#, the last tone to be derived in this scale. But a difficulty arises if we wish to repeat the order of the modern diatonic scale, starting with any other than the key-tone—a difficulty which necessitates the introduction of a new set of tones. These are the "chromatics" or "accidentals," indicated by the black keys which divide each whole-tone interval of the piano keyboard. There are but five to the octave, yet if the harmonic ratios were exactly held, it would be necessary to have countless additional ones. It has been computed that even the most important of the different tones requisite to accommodate all the possible musical intervals within a single octave would number above forty, and an instrument so constructed would be far too cumbersome for practical purposes. Besides, the expedient would be quite superfluous, since most of these finer nuances of musical purity go undetected; not because of musical deficiency on the part of the hearer, but because of his capacity to make an illusory correction in what he hears (*38*). Intent upon a right relationship of tones, he allows slightly incorrect intonations to pass unnoticed, as an interested reader will often fail to detect a trifling typographical error.

The introduction of five additional tones, so distributed within the octave that each whole-tone interval is virtually cut in two, answers all the requirements of free modulation, without occasioning any serious disturbance in the mind of the hearer. Following this principle of determining intervals, the piano-tuner is chiefly concerned that all the twelve intervals of the octave shall be *equal*. When this has been accomplished, we have the scale of "equal temperament" found on any piano or other musical instrument in which the tones of the scale are fixed independently of the musician's control. The mathematical formula for the vibration frequencies of such a scale is as follows:

$$n, \ \sqrt[12]{2}n, \ \sqrt[12]{2^2}n, \ \sqrt[12]{2^3}n \ldots \ldots \ldots \ldots \sqrt[12]{2^{12}}n.$$

Here the last term, of course, equals $2n$ which is the octave of n.

As a concrete illustration, the vibrational frequencies for this scale on a c of 512 v.d. would be approximately as follows:

c = 512 v.d.	g = 767.1 v.d.
c# = 542.4 "	g# = 812.8 "
d = 574.7 "	a = 861.1 "
d# = 608.9 "	a# = 912.3 "
e = 645.1 "	b = 966.5 "
f = 683.4 "	c¹ = 1024. "
f# = 724.1 "	

Here the ratio between adjacent tones is constant—1:1.059. Were it the musically correct interval of the semitone, the ratio would be 1:1.0666. In other words, the interval has been slightly "flatted," that is decreased, in order to gain an easy method of modulation.

Although the discovery and practical application of the tempered scale was brought about through the needs of the pianoforte as a musical instrument, this modification proves to be something more than a compromise measure. The piano-tuner does not resort to physical instruments to test the equal temperament of his intervals—he trusts his ear. In other words, he relies on his sense of proportion in testing the approximate equality of his intervals. Thus, while the purity of the harmonic intervals has suffered, the more fundamental and primitive sense of equal divisions has been restored to its place in the musical order. The pattern of musical thought does not altogether consist in tonic trends, tonal cadences, and a return to the point of departure; it also embraces elaborate proportional divisions, and the multiplications of a standard interval, whether this interval be harmonically "true" or not. Accordingly, the tempered scale of twelve equal intervals furnishes us with a very resourceful medium. While closely approximating all the significant harmonics, it also conserves the principle of equal division. In this way, it permits the free modulation of harmonic intervals, and at the same time provides for the identification and use of a variety of equal intervals—one-twelfth, one-sixth, one-fourth, one-third, and one-half the octave.

ORIENTAL SCALES

The acceptance of the law of equal division and its numerous musical implications is beginning to show forth in modern composition, although many "orthodox" musicians are still sceptical of its validity. It is less from the practice of modern musicians than from the investigations of the scientist in music, however, that we obtain our knowledge concerning the musical utility of this principle. The researches of Ellis (*35, 514*), Stumpf (*117, 127*), and others, have demonstrated that an original sense of proportion must have determined the formation of many of the scales employed by non-Europeans. In particular, the scales of the Javanese and the Siamese are evidential, for both peoples employ scales of equally-tempered intervals. The Javanese scale has six tones and five equal intervals to the octave, while the Siamese have a scale of eight tones and seven equal intervals. In explanation of these scales it has been suggested that they may have originated in the employment of a method which measured or divided strings, pipes, and percussion-instruments, such as the xylophone and the metallophone, into equal parts. In the case of the xylophone, if the various tones within an octave were obtained from bars of wood which varied by a constant increment of length from the one sounding the lowest tone to another sounding its octave above, we should, indeed, have a scale of equal temperament. But Stumpf has demonstrated that the accuracy with which these instruments of the East are tuned permits of no other interpretation than that of an original sense for equal intervals (*127*). Not only is it impossible that the musicians should have had access to the physical apparatus and mathematical devices necessary to determine the precise equality of the intervals these instruments are found to possess, but the Eastern xylophones also bear unmistakable evidence of having been whittled away or weighted with wax so as to alter their pitch and bring them into tune.

The music of the Orient affords many examples of theoretical scales comprising intervals of varied sorts and numbers within the octave; but as yet only the scales of Java
and Siam have been clearly shown to be scales of equal
temperament. Myers, however, finds an approximation
to a six-interval scale of equal temperament used in the
songs of the Malu tribal ceremonies (76). An analysis of
these songs, which was made from phonographic records,
showed two marked characteristics: first, descending intervals approximating the whole tone of our tempered scale,
and secondly, ascending octaves and fifths. In addition, intervals approximating a fourth also appeared, but never
any thirds or sixths. The antiquity of this music is indicated by the words employed, many of which are so archaic
that they no longer have any meaning for the singers.
Musically, the interesting features of these songs are the
apparent naturalness of the whole-tone intervals, and the
exact memory of the vocalist for absolute pitch, which
enables him to return to the original tone of the melody
after he has proceeded away from it as far as a fifth or
even an octave. The intervals, as sung, were not always
exact, nor could they be, for the whole-tone scale of six
equal steps contains no intervals that closely approximate
either the fourth or the fifth. Yet Myers appears to think
that the Malu music involves both a sense of equal intervals and a sense of harmony, the first being found in the
usage of equal steps approximating a whole tone; while
the second is indicated by the tendency of larger intervals to

approach the harmonic octave, fifth, and fourth. Since it involves no definite framework of an octave, this very primitive modus can hardly be said to imply a musical scale, yet it does suggest the influence both of the law of harmony and of the law of equal division. These two are combined, however, in a manner strangely incompatible, since the harmonic intervals of the fourth and fifth which the Malus employ do not occur in the whole-tone scale, whereas the third, which does, is apparently never used by them.

Quarter-tone scales are met with in the music of Arabia, Syria, and India; yet it is doubtful if these small intervals are ever freely accepted, like the larger intervals, as units of division; for while larger intervals are often augmented or diminished by a quarter-tone, this is not the same as using the smaller interval as an independent musical unit. Since the quarter-tone step is very near the threshold of volume-difference upon which the sense of interval depends, Eastern musicians, if they are able to recognize successive quarter-tone steps as true intervals, must be more sensitive to this difference than are their Western brothers. But, after all, a scale is only an ordered list of tones originally selected for a melodic purpose, and no melody is confined to single-step intervals. Hence so-called "quarter-tone music" does not necessarily imply that quarter-tone intervals are employed as they occur *seriatim* in the scale; indeed, it may mean nothing more than that some of the larger intervals differ from our standards by an increment or a decrement that approximates a quarter-tone (*40*).

But let us not be dogmatic. If the threshold for volume-difference were lowered, there would be an increase in the range of permissible intervals, and we are in no position to deny this accomplishment to the musicians of certain Oriental peoples. Quarter-tone music is of ancient lineage, as has already been shown with reference to the Greek "en-harmonics," and it may not be too wild a guess if we should generalize from a suggestion previously made that these small intervals were originally introduced in an attempt to record the slight changes of pitch-brightness which accompany inflections of the voice. Recently the eminent pianist Busoni has advocated the employment of third-tones in music, and he announces that he has accustomed his ear to tripartite tones "as wholly independent intervals with a pronounced character, and not to be confused with ill-tuned semitones" (*19*, 31). He thinks it probable, too, that "even sixth-tones will at some time be adopted into musical speech" (*ibid.*), and goes so far as to suggest an appropriate use to which they might be put. Max F. Meyer (*69*) like-wise accepts the smaller intervals as musical, and has constructed a quarter-tone scale of twenty-four intervals upon which transcriptions of Japanese melodies may be executed; but his is not a tempered scale of equal intervals. According to Meyer's view, this scale is merely a part of the "complete musical scale" which provides for all possible musical relationships, in the harmonic sense, that may exist between any two tones. In the "complete musical scale" smaller intervals like those advocated by Busoni do not appear as directly related tones, but they may, nevertheless, be employed as transitional intervals to attain more complicated harmonic effects than are commonly sought by Occidental musicians.

It is likely that the principles of equal division and harmonic sequence have each contributed to the standardization of these rarer intervals. In the absence of definitely-tuned instruments, such as the metallophone, the xylophone, and some wind-instruments like the Pan-pipes, it is difficult to ascertain when these principles are involved singly, and when a compromise has been effected between the two, as in our own tempered scale. The records of vocal performances are always highly untrustworthy, because of the variability of the individual performance. The interesting question as to the musical employment of intervals smaller than a semitone is therefore still open to debate and to further investigation.

THE WHOLE-TONE SCALE

As an instance of a compromise between equal and harmonic intervals we may cite the well-known Debussy "whole-tone scale"—the scale c, d, e, f♯, g♯, a♯, c¹, a scale having six intervals of two semitones each. On the tempered pianoforte this is the identical scale attributed by Myers to the Malu musicians, but as used by Debussy with harmonic setting it is not a pure scale of equal intervals. The weird effect of this music rests, in large measure, upon what appear to be strange modulations from major to minor modes. Being adapted to both these modes, which are basic

in our musical practice, we find it impossible to throw off
the influences of the tonic and atonic trends which they pre-
scribe.   But whoever will accustom his ear to take delight
in these unusual sequences may in so doing fall back upon
his primitive sense of equal division, and thus learn to accept the melody entirely freed of harmonic domination.   Nor

Example of Debussy's employment of the whole-tone scale

are the possibilities of equal division within our own twelve-
step tempered scale limited to the Debussy innovation.   The
Futurists in music are now contending for "free transposa-
bility" among all the twelve keys of the scale; there is no
logical reason why a chromatic mode of this sort should not
be available if only we chose to employ it.

This extension of musical thought will be referred to at
greater length when, in the following chapters, we undertake to consider the two chief forms of auditory expression
and communication, language and music.

CHAPTER IX

LANGUAGE [1]

EXPRESSION

THE part played by hearing in language is of great importance. As the term itself indicates, language implies the use
of the tongue in giving utterance to sounds; although there
are, of course, other forms of language which appeal to sight
and touch. Both gestures and the written page communicate
to us by sight, while the raised types of Braille make reading
possible through the finger-tips of the blind. Yet in tracing
the origins of communication, the essential function of language, we find that the utterance of sounds has played a
leading part in the development of a series of symbols to
represent man's common impulses and ideas.

Communication arises in the course of expressive responses
which are the chief organic functions of living beings. Upon
stimulation, the organism responds, or reacts, by an adjustment of some sort, and this adjustment is indicative both
of an original nature, and of the effects of experience in so
far as the organism has undergone modification through
previous adjustments. The response of the vocal cords is
not directly adaptive. The squeaking, squalling, crying, and

[1] For a more comprehensive study of the psychology of language see
Wundt's *Völkerpsychologie* (*146*) and Jespersen's *Language* (New
York, 1922).

bellowing utterances of animals are, in the first instance, but by-products of an agitation, more or less general, which leads the animal to express itself in numerous ways just because it can do so; although it may not always thereby satisfy its wants. But the utterance of sounds has, nevertheless, a survival value; for sounds possess a unique carrying-power, and can thus excite responses in other animals at a distance. So it may happen that relief comes to a suffering animal, or that other animals are summoned to share in a feast. Even an unintentional vocal utterance can result in circumstances beneficial both to the one uttering the cry and to those that hear it. The carrying-power of sound in the dark, and into distances where the recipient could not otherwise be affected by sight or touch, stamps it as the most useful of all organic means of communication.

COMMUNICATION

Communication, however, is something more than expression; for while the latter is the immediate and natural response of the organism, communication involves an ability on the part of the recipient to understand what the expression signifies. We all react expressively in numerous ways that are not communicated to others; but among all our expressions those of language constitute the most important avenue of communication, because these are utterances that have become standardized with respect to meaning. In the slow process of standardization, sound-utterances have become the symbols of meaning for facts, adjustments, motives, impulses, and ideas.

It is not within our province to study the origins of language, or the development of its symbolism, for we are here concerned only with that aspect of language which is vocal—the utterance and perception of a sound for purposes of communication. Our problem, therefore, is limited to a consideration of speech as an instrument providing sounds of sufficient range and discreteness to be of use as symbols that can be related in meaningful communications. Unlike musical tones, the relational system of linguistic sounds is not completely determined by the characteristics of the sounds themselves. Tone, with its elaborate system of tonality, has no place in spoken discourse. The assistance music may render to language is largely accidental, or, at best, an aid to memorization; even here it is the rhythms and assonances, rather than the tonal relationships, that account for the easily-memorized ballad-literature with its so-called melody. Apart from the affective nuances of experience, a thought is neither more readily communicated, nor better understood, when the medium of its expression is song instead of speech. Indeed, the presence of a definite melody attached to a communicative utterance is confusing to most listeners of average musical ability. The words of a song are not a musical asset so much as a compliance with the customary usage of the voice. If one could construct an arbitrary set of voice-tones that would represent maximal tonal values, the purely musical effect of singing might even be enhanced. Since all the sounds of the voice with which we are familiar are the ones used in linguistic communication, it is quite natural to employ these in musical vocalization. Possibly, too, the human voice is incapable of any musical effects more beautiful than those provided in conjunction with the series of vowel-sounds.

They who demand a clear understanding of all the words of a song are either persons whose musical interest is subordinate to their interest in the linguistic communication, or musicians whose ability to comprehend musical expression is so great that they can readily compass in one unitary impression the quite separate arts of poetry and music. In the work of Richard Wagner an attempt was made to fuse all the arts: plastic, decorative, dramatic, poetic, and musical; yet no one doubts that it is the musical rather than the dramatic, the poetic, or the pictorial aspects of the music-dramas to which we attach the greatest worth.

PHONETICS

Language deals with symbols—conventions valuable not alone in themselves, but also for the meanings they give rise to. Since regard must be had for individual capacities both of utterance and perception, the sounds employed are far from being arbitrarily chosen. In order to be effective, linguistic sounds must be uttered with facility, and apprehended with ease; differentiations of sound must be clear-cut both in their production and in their perception. The vocal instrument follows certain laws of production best understood from a study of phonetics. The more or less uniform capacities of the vocal cords as generators, and of the mouth-, nose-, and throat-cavities, aided by the tongue, as resonators of sound, furnish the basis of vocal utterance.

The two kinds of sound thus generated are the vowels and the consonants, which we have already described. A number of the consonants are explosives, dental or labial, such as b, d, g, k, p, t, and also sometimes j and v. Certain regions of pitch afford characteristic formants for all the other consonants, as has already been indicated by the discussion of Stumpf's analyses. The science of phonetics has undertaken to study the genesis of vocal sounds with reference to the physiological adjustments of the vocal organs. There has been a considerable amount of controversy as to the relative importance of acoustics and physiology in phonetics, but the disagreement among earlier investigators in the assignment of pitch and resonance to the vowels, and the difficulties of analyzing the noisy constituents of the consonants, have led phoneticians to favor a physiological study of throat-, tongue-, lip-, and jaw-positions as the best means of solving their problems (44, 65). Since we now have at our disposal the exact methods of Stumpf and Miller with which to analyze the vocal sounds, we may in future expect a union of the results thus obtained with the genetic method of approach that has generally prevailed in the past. In consequence, a more complete science of phonetics is promised than has hitherto been possible; for both the motives

and the functions of articulated sounds will be clarified
when the physical and the psychological nature of these
sounds can be scientifically defined. While it is not within
our province to bring these data, so far as they are yet
known, into connection with the physiology of vocal production, certainly the phonetician must grapple with this
problem. For our immediate purpose it may be sufficient
to remark that, while modifications in the usage of vowels
and consonants are met with in different languages, the more
essential features are universal, and point to the influence
of a common vocal capacity which determines what sounds
shall be adopted for linguistic purposes.

The character of the sounds employed in language is not
merely a matter of vowels and consonants, for rhythm and
rhyme, assonance and inflection, accent and cadence, are
also important in determining sequences with differential
features. So far as concerns communication by definite and
unambiguous auditory symbols, a marked contrast in sounds
is eminently effective. But we must also reckon with facility
of expression and facility of sequential apprehension. As
for the first, the vocal apparatus must be able readily to ad-
just itself to its task in order that speech may be fluent. The
harshness of contrasting sounds is subdued by a tendency
to employ intermediate vocalizations as we pass from one
to another. These transitions lead to slurring, and also to

the mediation of vowel-sounds in their appropriate order. It is easier, for instance, to utter the vowels, Continental fashion, in the order *u, o, a, e, i,* than in the order of their occurrence in our alphabet, *a, e, i, o, u.* In the first order we have a simple transition from low to high formants; whereas in the second we start from a middle position, and pass to the highest, then return to a lower register, and end with the lowest. The nature of the vocalic sounds employed also contributes to the facility and to the æsthetic quality of their effects. In the case of Whitfield's famous word of beauty, *Mesopotamia,* the transitions are easily made, and, except for the short *e* at the beginning, the vowels are all full and definite. By way of contrast, we find in a word like *disaggregate* a harsh combination of sounds, in which the vocalic transitions are difficult.

Our perception of words is also influenced by transitions of volume, pitch, duration, intensity, and brightness. The pleasing quality of a vocal utterance is not explained by striking contrasts of sound, but by the capacities of vocalization and of orderly apprehension. The smooth fluency of spoken or written language which we term *style* has among its elements rhythm, the repetition of like sounds, quantity or duration, stress or accent, and gradations of volume and pitch; all of which are subject to the habitual capacities of vocalization and perception.

The rhythm of prose, though not so definite as the rhythm of verse, is, nevertheless, discernible. No sequential expressions of a like nature can avoid rhythmical form, either simple or complex. In both verse and prose there is an underlying rhythm attributable to a fundamental "sense" of time, which in reality is a perceptual elaboration, involving both organic and kinæsthetic sensations. One might say that the attitude toward sequential sounds, whether heard or uttered, always involves a temporal or rhythmical pattern upon which the syllabic sound-structure of the verse or prose is, as it were, superimposed.

In his recent study of the rhythm of prose, W. M. Patterson (*85*) gives an ingenious interpretation of the phenomena. After distinguishing the individual characteristics of persons who are "aggressively rhythmical" in their attitude from those who are not, this writer contends that the essential difference between the rhythms of prose and verse is that the rhythmical pattern of prose is syncopated, while that of verse is produced by coincidence between a *ground-rhythm* and the syllabic sound-structure. In prose, the time-intervals established between the fundamental time-sense and the syllabic structure are irregular, and the beats are marked by the varied pattern of a perceived movement. Since the emphasis of beat may be made apparent by either pitch, stress, or duration, the substitution of one of these elements for another makes the rhythm of prose a matter of syncopation. "Spontaneous substitution," writes the author, "means here simply: first, that when two sets of time-intervals are occurring concomitantly, one a unit-series and the other a more complicated arrangement (to be co-ordinated with these units and then organized), the unit-intervals, in their successive occurrence, establish the fundamental continuous experience of rhythm; secondly, that this experience is not disturbed, but rather made more interesting, by the fact that any one of the units can be accompanied by two, three, four, etc., sub-divisions of a unit-interval, produced by the beats of the more complicated series" (*ibid.*, 71). In verse, on the other hand, one seeks a more direct correspondence of beat with beat in the two series. Thus metrical form is regulated to the underlying time-unit interval, and syncopation is not necessary in establishing these correspondences. In either case the attributes of sound contribute to the stress or accent, and likewise to the duration, which together constitute the complicated rhythmical order that is produced. But so natural is the proceeding whereby we bind the sounds as uttered or perceived to the fundamental unit-intervals of our time-sense, that accents are readily supplied wherever they are not provided by the sounds themselves. We also tend to dwell upon and to repeat sounds the vocality of which is pronounced. Effectiveness in production, and clearness of impression, lead us to regard certain sounds with a special interest, and it is the sensitive ear and the delicately-adjusted organ of speech that register the effect of *le mot juste*. Likewise an "aggressive timer" is able to organize into a rhythmical pattern even the loose structure of ordinary prose. Lacking some measure of "aggressiveness," it is impossible for one to apprehend the complete rhythmical effect even of verse. In music, one main difference between satisfactory and unsatisfactory execution lies in the ability of the performer to bind his measures into a single onward surging movement, so as to sway his hearers without the let and hindrance of an unwelcome pause or break in the rhythmical pattern. Similarly the skilled speaker plays upon the susceptibilities of his audience by the balanced rhythm of his utterances.

Individual differences count heavily in the determination of rhythm, and in the demand for rhythmical expression; and hence some of us are poets, while others are not. Yet the transition from prose to verse is never an abrupt one, and, as Patterson tells us, the unsymmetrical prose of ordinary conversation is readily distinguished from the grouped symmetries of, say, Biblical prose. Likewise we can distinguish grouped symmetries from the exact correspondence of syllable and beat which characterizes poetry, and also from the grouped though unsymmetrical phrases of "free verse." According to Patterson's definition, it is not the rhyme and the equality of metrical "feet" that define poetry, but the coincidence of definite sounds with definite time-intervals. To be sure, gaps may occur in the series of sounds; but it should be remembered that the pause is a durational element, and hence the *cæsura* is no less real as a structural moment of the sequence than is any other vocal unit.

CADENCE

In addition to the effects of rhyme and rhythm, those of rising and falling inflections are also important. The rich resonance of the bass stands apart from the high, shrill treble. Here again we employ transitions in the volumic mass and pitch-brightness of sounds to procure effects that are both pleasing and serviceable in communication. Indeed, the conventions of a language which are dependent upon these effects are so numerous and so involved as to be almost past understanding. Although by no means arbitrary in origin, their evolution has been a slow process of blindly groping experiment, in consequence of which we are often at a loss to trace the genesis of these usages. Yet certain of these conventions indicate clearly enough that a natural selection has ever been their guiding influence. The falling inflection at the end of a sentence is an indication of finality through relaxation; the rising inflection of the question is typical of the tension in the mind of the questioner; and the employment of intensive emphasis is obvious enough either at the beginning of an utterance or as an exclamation.

Attempts at the construction of a universal language have always met with practical failure, because no artificial language can meet these natural though obscure requirements of utterance and apprehension. Logically, the artificial product may possess distinct advantages over any tongue of natural origin; for nature does not work with scientific accuracy, and her achievements are often hard to rationalize. Yet whatever she produces is in some way suited to the organs of its production, and so long as we lack a comprehensive knowledge of all these subtle distinctions and interrelations of functional efficiency, no artificial product, however ingenious, or however nearly it may approach to the nicety and exactness of a mathematical formula, can possibly prove acceptable in lieu of one's native tongue, with its natural adaptation to the requirements of expression and communication.

THE SCIENCE OF PROSODY

Numerous contributions have been made to a science of prosody, among which the well-known work of Sidney Lanier (55) may be mentioned, because the plan of his ingenious study so nicely accords with the data we have set forth in this volume. Lanier starts with a definition of "verse" as a group of specially related sounds. These are the sounds of speech interspersed with silences, which together constitute the rhythmical structure of a sequential auditory movement. The elements or "particulars" which he recognizes in the structure of a poem are four: *duration, intensity, pitch,* and *tone-color* or timbre. Of these, all but intensity, he thinks, are capable of exact co-ordination by the hearer. Thus the co-ordinations of duration give us rhythm, those of pitch give us the melody of speech, and those of tone-color, the rhymes, assonances, alliterations, etc., of prosody.

In agreement with Patterson, our fundamental "time-sense" provides a rhythm of the first order with respect to the relative duration of each primary unit or syllable of speech. The *quantity* of these units, while it is not limited to a long and a short in the proportion of two to one, as indicated by classical prosody, always approximates some definite proportion, such as two to one, three to one, four to one, etc. Lanier does not seem to have thought of the possibility of other simple proportions such as three to two, or four to three, yet these, as well as equal beats of one to one, merit consideration. The problem of quantity as the basis of rhythmical co-ordination is further complicated when one examines the auditory constituents of the sound-unit more closely than Lanier was able to do. While a fundamental "time-sense" may employ units in a sequential utterance, the units themselves are defined by virtue of pitch, brightness, and volume, as well as by intensity. The unit is a complicated integration to begin with, but it suffers a modification whenever it enters into an integration with succeeding units to form a rhythmical structure. Sterzinger (*113*) has shown that if the time-interval and the intensity are held constant between successive sounds, the accentuation may be one of quantity attaching to the more voluminous of the sounds, or it may be one of pitch-distance in which the higher pitch usually carries the accent.

On the basis of accentuation—either subjectively indicated by the relative importance which is attached to these structural details, or objectively compelling, as when the time-intervals or intensities set the measure—groups are formed, as musical bars or metrical feet. This is the second order of rhythm proper to which Lanier refers. A third gives us the *phrase,* as an alliterative or emphatic word-group, which may modify the even tenor of the rhythm in order to accen-

tuate the meaning of certain words. A fourth order is the
line or *metre;* a fifth is the *stanza;* while a sixth, and last, is
the complete poem.

It is obvious that the time-relations of spoken and perceived words will vary greatly from individual to individual;
yet despite the irregularities of accent that may be introduced by attitudinal variation, definite limitations are set,
within which the rhythm as a whole must progress. Study-
ing the simple rhythm of musical intervals in groups of two
or three tones, Sterzinger found certain of his observers predisposed to favor volume, others pitch-distance. Yet with
practice all tended towards a mean type in which a reconciliation of the two was effected. With respect to the accents of pronunciation and of logical derivation, which
Lanier considers to be irregular, these are not truly disruptive of the fundamental rhythm, but more in the nature
of syncopations and transitions of tempo.

As regards the "tunes" of verse and prose, it is not primarily musical intervals that are responsible for these cadences, but the integrative effects of brightness, volume, and
intensity. While the vocal formants are voiced upon fundamentals that may have a definite pitch, the total structure
of a vocable is not that of a musical tone. Hence the melody
of speech is an effect of cadence in which the salient pitch
essential to a musical tone is lacking. In the passage from
one vocable to another we are more apt to have a *portamento* glide than a clear-cut step such as characterizes the
musical interval; and while a more or less definite distance

is traversed in the auditory scale, and the syllables of speech are themselves set off as units of varying size or volume, it is not musical tones, but vocables, which mark these steps, and define their units of sound. The effect of a speech-scale is produced by variations of regional pitch and volume. While the tunes of verse are more or less conventional cadences, they are not melodies in a strictly musical sense. It is not unnatural that a vocal cadence should often approximate a definite musical interval, and the transition from speech to song is not abrupt, but gradual. Yet if we measure cadences in terms of the pitch of the voice, we find many individual variations in the pronunciation of the same phrase, which would be intolerable to our ears if the phrase were supposed to be sung with the same melody. Even when a certain speech-tune is so conventionalized as to compel a like enunciation by different speakers, the syllabic units will vary, because only a regional pitch is requisite to voice the prescribed formants of the vocal structure; and hence both the distances between vocables, and the volumes of the vocalic structures themselves, are subject to individual variations such as music with its exact, salient pitches does not permit.

The essential features of a melodic progression can be studied with simplified sound-structures, namely, the pure tones, each of which has a salient pitch and a definite volume in accordance with which the intervals of the progression can be neatly defined. A speech-tune, on the contrary, can never be so simplified. While the formants of the vowels and consonants may be considered as simple sound-structures, they involve a considerable amount of resonance even when whispered; and when voiced, the fundamental tone upon which they are uttered is their most striking phenomenal constituent. Yet, as we know, different vowels and consonants can be voiced upon the same fundamental. The formant introduces a nuance, or color, to the sound-structure which renders it vocalic; but when heard alone, the formant may be virtually imperceptible. As Stumpf has shown, the primary formants for the vowels *ü*, *e*, and *i* are so close together that without the accompaniment of their secondary formants, which lie lower in the scale, they are quite indistinguishable.

Our conclusion is that the tunes of speech are not based upon musical intervals, but upon cadence, the syllabic units of structure being always complex sounds which for the most part glide from one region of pitch to another. For this reason, variations of brightness are more significant in the integration of vocal units than variations of pitch. Indeed, the whole movement of speech is much more flexible than can ever be the case in a musical melody. Yet the movement is by no means arbitrary, since the formants that characterize the vocables have definite structures for which a definite regional resonance must be allowed.

The child learning to read pronounces the monosyllabic words of his primer: "I-see-a-cat-I-see-a-rat," with little or no tune, because he is so intent upon the articulation of each word that he fails to formulate a structural cadence out of his sequence of utterances. There may be some evidence of a primary rhythm, but there is no movement in which the units of sound are gathered together in rises and cadences. At the other extreme we have the sing-song and the chant, in which the tonal constituents of speech are emphasized, and more or less definite intervals of a musical order introduced. Thus the mediæval formulæ for Church lessons, collects, etc., is exemplified in the following well-known Gregorian chant.

Sic can - ta com - ma, Sic du - o punc - ta: Sic ve - ro punc-tum.

Sic sig-num in - ter - ro - ga - ti - o - nis?

The third section of Lanier's *Science of English Verse* deals with the color, or timbre, of speech, which concerns the order and arrangement of vowels and consonants in the construction of rhymes, and likewise the repetition and variation of vowels and consonants. The structure of phrases, lines, and stanzas owes much to alliteration and assonance, as well as to the effects of transitions and contrasts of timbre. A refined sensibility to these inner connections and disruptions is always characteristic of the literary artist in the composition of prose and verse, whether intended for the printed page or for oratorical presentation. We need not enter into the details of these structural effects, good and bad, but the following illustration from Lanier (*ibid.*, 302) will indicate the intolerable effect of an exaggerated utterance of the same vowel-color.

> Tis May-day gay: wide-smiling skies shine bright,
> Through whose true blue cuckoos do woo anew
> The tender spring, etc.

In the first four words of the first line, the vowel-color *ay* recurs three times consecutively; in the five next words of the same line the vowel-color *ī* occurs five times nearly in succession, the only break being the vowel-color *ĭ* in "ing."

In the second line, the vowel-color *u* (long *u*, or *oo*) occurs eight times, relieved only by the shorter sound of *u* in "cuck" and the color of *a* in "anew."

Not only the vowels, but the consonants as well, play an important part in the coloration of speech. As instances of the succession of consonant colors we may quote the following passages, again from Lanier:

The daily torment of untruth (from one of Daniel's sonnets) where the ear may not only co-ordinate the alliterative *t*'s which begin the accented syllables "tor-" and "-truth," but may take further account of the *d* in "daily"—for *d* belongs to the class of T-sounds . . . occurring closely together; and this is a syzygy [1] of T-sounds.

Similarly let the student pick out the *m*-colors, and the *d*- and *t*-colors, in the opening of Shakespeare's sonnet.

> Let me not to the marriage of true minds
> Admit impediments;

[1] From *sunzugia,* yoking together.

these constituting respectively syzygies of *m*-sounds and of *t*-sounds.

These illustrations will serve to indicate a few of the elements of vocal timbre which supply important addenda to the rhythms and the tunes of speech and versification. Taken together, they enable us to comprehend the principles of sound upon which language becomes elegant, and to which belles-lettres owe their existence.

THE PROBLEM OF LINGUISTIC SOUND

Although our knowledge of linguistic sound is still imperfect, progress toward enlightenment has been máde in the study of prosody, in phonetics, in comparative linguistics, and in anthropology. The problem of speech has often seemed obscure, but is by no means beyond solution. With knowledge already accessible, or not too difficult of acquisition, it should be possible to discern the trends of linguistic evolution, and thus, by appropriate educational methods, to guide nature in a more effective and economic use of her agencies. Systematic attempts of this sort have frequently been made with reference to written language in the way of simplified spelling. Though the pedagogy of oral language is not so well-developed, few teachers of language now attempt to introduce a student to a foreign tongue without recourse to the phonetic principles of pronunciation. To know the proper placing of vowels and consonants in the buccal cavity is of the greatest aid in approaching the utterance of foreign sounds. In this connection the study of the vowels and consonants as elements of speech is especially important, because we now have a considerable amount of precise information regarding the nature of the vocal formants, each of which requires a special adjustment of the mouth and throat for its utterance.

It appears that vowel-sounds are normally some ten times longer in duration than are the consonants (*60*). In certain cases of deafness this greater duration is indicated in that the vowels can be heard by the patient after the hearing of consonants has been lost, which suggests the wisdom of teaching the consonants along with the vowels, instead of using a method of drilling the student in the separate sounds; for the consonant is in its natural setting when attached to a vowel, whose longer duration carries it better than when it is sounded alone. If the sound is articulated in this way, peculiarities of utterance come into prominence which would otherwise be difficult clearly to apprehend.

The first problem of language, in so far as it concerns our present interest, is that of analyzing the complex noises and vocables employed in oral expression. Upon the basis of analysis one might proceed to infer principles of orderly arrangement with reference to the attributes of sound, and thus to determine the influence of each attribute, both singly and in combination, and its effect in the production of simultaneous and sequential patterns of expression and communication. Rhythm, assonance, cadence, the proportions of volumic masses, and the law of return has each its part in determining vocal utterances that are suited to auditory perception. But great importance likewise attaches to the method and capacity of vocalization in the individual. Here one must resort to the physiological principles of phonetics.

With this bare program we shall be content, for significant research has only begun, and its ends are as yet too dimly descried to warrant useful speculation as to its outcome with respect to the best methods of linguistic study. We therefore turn again to a field in which our knowledge is somewhat greater—to music as a mode of thought, of expression, and of communication.

CHAPTER X

MUSIC

THE ELUSIVE NATURE OF MUSIC

AMONG the products of man's genius none offers a more fascinating problem than music. Rooted though it is in that abundant experience which has been our heritage and is our being, music seems a thing apart, putting forth its bloom in a world peculiarly its own. Schopenhauer, who recognized this fact more clearly than do most thinkers, remarked: "Whoever gives himself up entirely to the impressions of a symphony seems to see all the possible events of life and the world take place in himself; yet if he reflects, he can find no likeness between music and the things that passed before his mind." Browning touched with even greater eloquence upon the mystery of musical art, in *Abt Vogler*.

> Had I painted the whole
> Why, there it had stood, to see, nor the process so wonder-worth.
> Had I written the same, made verse—still, effect proceeds from
>   cause.
> Ye know why the forms are fair, ye hear how the tale is told;
> It is all triumphant art, but art in obedience to laws,
> Painter and poet are proud in the artist-list enrolled:—
> But here is the finger of God, a flash of the will that can,
> Existent behind all laws, that made them, and, lo, they are!

And I know not if, save in this, such gift be allowed to man,
That out of three sounds he frame, not a fourth sound, but a star.

But we must leave to poets the wonder, for to the man of science neither the fundations of this creation nor the plan of its structure is mysterious. The artist in music, living in a world of tone, is content to take his material as he finds it. It is no concern of his to inquire into origins, and as for his compositions, constant trial and experiment are the means whereby he finds what can, and what can not, be accomplished by combining tones. For critic and theorist, however, this empirical procedure will not suffice. The basis of true criticism must be more firmly laid, and a theoretical interest naturally arises in regard to every different mode of human expression. What, then, are the origins of music, and what are the peculiarities that distinguish the patterns of musical art?

We have already seen that music is an expression and a communication behind which there exist acts of thought—*musical* thought. As Combarieu has written, *"La musique est l'art de penser avec des sons"* (20). The materials—the contents—of music are tones. The principles of thought whereby these are combined in expression and for communication are the ones we have ascribed to tonality, and the elements of the matter have already been laid before the reader in the preceding chapters of this book. What remains for us now is to add certain details regarding the development of music as an art.

MELODY

The simplest kind of music is melody; and a melody is a sequence of tones developing an orderly theme to express a musical thought. Its analogy may be found in the verbal sentence; for music, likewise, has its simple phrases, its antecedents or questions, and its consequents or answers. These are the *periods* of a musical sentence (*111,* 50 f.), and they may be grouped together, as are the paragraphs, sections, and chapters of a verbal composition.

It is not difficult to comprehend the principles of melodic composition. Indeed, we have already learned what these are, and can now apply them to concrete instances. Let us consider, for example, the music of the Siamese. The distinguishing feature of their melodies is the complete absence of a tonic, since the Siamese scale affords no musical relations in which a tonic appears. Aside from this, a Siamese melody is not obviously different from any other; the law of return and the law of cadence are both observed in its musical patterns.

FROM THE FAN DANCE II

Song ♩=88

Orchestra unisono without embellishment

(or)

ritard

Specimen of Siamese Music

Instead of the dominating sway of the tonic, this Oriental music permits a freer and far simpler usage of recurrent intervals made up of one, two, three, or more, equal steps. Since there is but one basic interval, the musical pattern is determined by a sense or discernment of a certain definite interval and its multiples. In employing a scale of equal intervals it is evident that the melody can be transposed from one place in the scale to another without altering its tonal relations in any way. The introduction of *accidentals* in order that the melody may be reproduced at a higher or lower level or pitch is quite unnecessary.

But along with the gain of a free range of transposition there goes the loss of all harmonic possibilities. Harmony, as we understand it, has no place in such a system, because harmony is a product of fusion, and, save for the octaves, there are no fusing intervals in this scale. A simultaneous duplication of the melody at different levels of pitch is possible, and to some extent practised; though, if it be regarded harmonically, the polyphonic effect is, of course, discordant, except in cases where the interval between the tones happens to be that of an octave. Yet the discrepancy between equal intervals and those that fuse in harmony is not always very great, and it is noteworthy that in both the Siamese and the Javanese scales the discrepancy is least in the case of those intervals which most closely correspond with the fourth and the fifth of our diatonic scale.

A closer study of the intervals of different scales is readily made by comparisons in terms of *cents*—a scheme of dividing the octave by hundredths ("cents"), so that its compass is 1,200 cents. In the *just* intonation of true harmony a semitone-interval will then have 112 cents, a small whole tone, 182 cents, and a larger whole tone, 204 cents. Adding together two semitones, two small tones, and three large tones, we have the sum of the octave, or 1,200 cents. In the case of the tempered scale of twelve equal semitones, the semitone is uniformly 100 cents, and the whole tone uniformly 200 cents. With reference to this rule, the following comparison is made between the intervals of the tempered scale and those of just intonation.

In tempered intonation:

c–d	c–e	c–f	c–g	c–a	c–b
200	400	500	700	900	1,100

In just intonation:

c–d	c–e	c–f	c–g	c–a	c–b
204	386	498	702	884	1,088

On the whole, it will be observed that the discrepancies between the tempered and the harmonic intervals are negligible; disagreement is greatest at the major sixth (c-a), where is 16 cents, and at the major third (c-e), where it is 14 cents; while the smallest differences are at the fourth and fifth (c-f and c-g)—each being but 2 cents.

PRIMITIVE MUSIC

Extending now our comparisons to the three non-European scales known to employ equal intervals, we meet with greater discrepancies, as might be expected. In the whole-tone or six-interval scale, the perfect fourths and fifths are wanting. Otherwise the tones are selected from our chromatic scale, and measure equal divisions of 200, 400, 600, 800, 1,000, and 1,200 cents from the key-tone. In the seven-interval scale of Siam, the fourth tone, or f, is represented by $514\frac{2}{7}$ cents, and the fifth tone, or g, by $685\frac{5}{7}$ cents; each of these being $16\frac{2}{7}$ cents "off" the corresponding interval of *just* harmonic intonation. Although these mistunings are not appreciably greater than some we meet in our tempered scale, they are more significant because they attach to tones which in our scale represent the harmonic fourth and fifth. In the five-interval scale of Java, the third tone has 480 cents, as compared with 498 for the harmonic fourth, while the next tone has 720 cents, as compared with 702 for the harmonic fifth. Here the disagreement in each case is 18 cents; though, again, it is not so great that one can not readily recognize the intervals as being, respectively, fourths and fifths. All the other intervals of the Siamese and Javanese scales show much greater differences from any corresponding intervals of the harmonic scale.

Although the music of Siam and Java follows the principle of equal intervals rather than that of harmonic divisions, the close approximations of the fourth and fifth which

appear in both these scales may yet indicate a harmonic influence in the selection of the total number of intervals in
an octave. In other words, when the octave is divided into
five and seven equal steps, intervals appear which approximate the true fourth and the true fifth. In the first case, the
fourth is flatted and the fifth sharped, and in the second
the fourth is sharped and the fifth flatted. An influence
leading to an approximation of these two important harmonic intervals may have been felt in the original tendency
to provide a scale of equal intervals, thus accounting for
the choice of five and seven intervals, rather than six, eight,
or nine—where no such approximation is possible (*134,*
131 f.).

It appears, then, that the harmonic usage of these two intervals is made possible in each of these scales with but a
slight deviation from their proper tuning. It would be unwise to stress this interpretation of the reasons for choosing
intervals of five and seven rather than some others, especially since it has been reported that polyphony as practised by non-European races is quite as prevalent with non-
harmonic intervals as with those that resemble the harmonics. Possibly the disposition to favor equal intervals is powerful enough to dominate even the simultaneous
employment of tones that do not fuse, thus giving rise to
a species of harmony quite at variance with the practices of
modern music. On the other hand, there is positive evidence that Oriental musicians favor intervals of the fourth
and the fifth, as well as the octave, both in two-part singing

and in instrumentation; it is therefore worth noting that the mistuning of the fourth and fifth is not very great either in the Siamese or Javanese scales. When singing, and when using instruments whose scales are not objectively fixed, it would be an easy matter, if the need were felt, to approximate the corresponding intervals of harmonic intonation lacking in these scales.

As regards the whole-tone Malu music, studied by C. S. Myers (76), we are unable to draw any very definite conclusions, because this music is vocal, and hence subject to variation of the intervals employed. But Myers found a decided tendency to employ equal whole-tone intervals of a descending order, and he also records descending fourths, and ascending octaves and fifths, although neither the true fourth nor the true fifth appears in the whole-tone scale. If we accept Myers' analysis, the Malu tribe possesses a sense of equal intervals, and also a sense of the three fundamental harmonic intervals. "Indeed," he writes, "I do not think that, after a careful study of the Malu music, any one can doubt that the octave, fifth, and fourth intervals therein employed have developed quite independently of the fusion-effects produced by such accords. It seems certain that the use of these intervals depends directly on the pleasure derived from the relationship between the two *consecutive* tones, and not on the fusion-effects obtained by hearing the tones *simultaneously*. As I have already insisted, the memory for absolute pitch has played an important part in furthering this relationship; for the intervals are employed under the precise conditions most unfavorable for the preservation of tone relationship, e.g., between the ends of almost chromatic passages, and between the ends of a slow *portamento* glide" (*76*, 579 f.).

Thus an original tendency to make use of the three chief musical intervals seems here to be indicated without reference either to fusion or to the employment of harmony. It must be noted, however, that in primitive melodies, carried solely by voices, the exigencies of a fixed scale are not felt. It may even be doubted if, under these conditions, the octave has the same significance in "framing" the melody that it has when cultural tradition has defined a scale, and musical instruments have been constructed to express its varied possibilities. The evolution of the Greek scale illustrates such a development; while in Malu music we seem to be confronted with musical tendencies that have not yet become definitely formulated. We need not presuppose an octave-setting for the use of descending whole-tone steps; and as for the larger intervals, the octave, the fifth, and the fourth, they merely indicate the employment of more or less standardized intervals of a semi-harmonic order. It may be worth emphasizing that Myers reports the fifths as ascending, and the fourths as descending, for both are contrary to the tonic trend, and hence do not imply a harmonic setting. As Myers himself has said, it is the law of return and the sense of absolute pitch, rather than harmonic or fusional effects, that seem to regulate this practice.

As to the general nature of the far-Eastern musical com-

position, it is essentially that of a melodic sequence, duplications of the melody occurring mainly by way of an ac-
commodation to the varied tonal range of different instruments, and especially of the human voice. This was likewise a common practice in the mediæval music of Europe.
Chinese symphonic structures seem to indicate an interesting evolution of melodic dominance when each instrument
takes the theme and elaborates it more or less independently.
Symphonies of this order are based upon a kind of free
counterpoint, in which the contrapuntal effects are not har-
moniously worked out. A definite plan is maintained in the
theme which characterizes the composition, yet no special
regard is paid to the fusion of tones that chance to fall together. It may require a very different musical attitude to
appreciate a symphony of this order, but there is no inherent
fallacy in the *rationale* of the procedure.

RHYTHMICAL ACCOMPANIMENT

Oriental melodies are not altogether lacking in accompaniment; what they fail to provide by way of harmony they
offer in rhythmical elaborations. We must remember that
music is always essentially a movement, or progression, in
time. The factor of duration is therefore a most important
adjunct, and while our own music has more and more tended
toward fixed and somewhat inelastic rhythmical forms or
measures, the music of the Orient, wanting the restraint of
a strict counterpoint and chordal harmony, has been free

to develop rhythms unknown in the music of the Western world. Not only are rhythms of an unusual type employed, such as measures of five and seven beats; but also transitions involving the most intricate of rhythmical patterns. The drum and other percussion-instruments are much used in the production of these effects, the subtlety of which baffles our uneducated powers of apprehension. Accentuations of tones and intervals serve to enrich and embellish the melody with a unique "rhythmic harmony"; yet so seldom is the music of these Eastern peoples reduced to written notation that our knowledge of its details has accumulated but slowly, and only in recent years with the aid of phonographic records have we been able to approach the problem in a scientific way. The ear of the occasional traveler has proved far too untrustworthy, for he is always predisposed to hear the intervals, and to feel the rhythms, with which he has become familiar through his own training.

HARMONIC INTERVAL

Though adequate information is lacking, it has commonly been stated that the music of East Indian tradition is based upon the varied use of a seven-tone scale, while that of the Chinese rests upon a five-tone scale; a difference which suggests that the Siamese may have followed the Indian tradition, while Javanese music may belong to the Chinese system (*84*, 21 f.). If this be the case, the scales of Siam and Java are truly typical of the most important musical traditions of the Orient. Since all these scales imply acceptance of the octave or some other harmonic interval as a foundation or framework, harmonic principles, though less evident, may not be discarded as irrelevant in the interpretation of Oriental music. On the contrary, as we study more carefully the music of Eastern culture, we may expect to find that the principles of harmony and of equal division have both been employed, although from our present knowledge it appears that the equal interval is the basic principle in all musical evolution.

Turning now to a consideration of our Western musical structure, we observe that the principle of harmony is paramount. Related intervals are the intervals of the harmonic series of partial tones; each tone of such a melody being set, as it were, in a chord. The æsthetic demand is not merely that of an ordered sequence, but likewise a demand for the accompaniment of harmonic or fusing tones to form an accord. This second requisite leads to harmonic elaboration; for simple melody is no longer the prime desideratum in the complicated interweaving of melodies found in modern polyphony. Under these conditions there also arise unique harmonic effects, in which the successive combinations of tone move through stages of discordant tension to concordant resolution, accompanied by massive effects of resonance that modify and supplant the simpler mode of melodic sequence.

Stumpf tells us that the basis both of music and of the scale is to be found in the major and minor chords (*121*).

Analysis shows these chords to consist of three tones each, the first chord embracing a major third and a perfect fifth, the second chord a minor third and a perfect fifth. These tones may be arranged in any order so that the total span of a triad is either a fifth or a sixth, and the internal intervals are both thirds and fourths.

That musical thought should be altogether subservient to this narrow interpretation of harmonic relationship is clearly inadmissable. Having accepted the principle of equal division as more fundamental than the principle of harmony, we can not suppose that harmony has entirely superseded equal intervals. Furthermore, a very important musical effect is produced by the use of adjacent tones; the succession of whole-tone and semitone intervals is a means of transition which not only bridges the gap of larger intervals, but also affords a ready modulation from one key to another. These intervals of propinquity, to be sure, are found in the harmonic series, yet the ease with which we accept a tempered scale that so reduces the semitone as to destroy the true tonic significance of the 15 to 16 ratio suggests that in point of fact we are often employing our sense of equal interval in making these transitions. Indeed, it is highly probable, as we have already remarked, that all intervals were originally based upon a sense of volumic proportions, and that from this foundation modifications were gradually introduced so as to bring about the *just* intonation necessary for harmonic progression with its peculiar tonic effects. Our *tempered* scale, though historically much later than the

scale of *just* intonation, could then be regarded as a "natural" scale, and accordingly one that provides us with
musical opportunities not to be found in the more precise
harmonic systems. With its adoption, we have undoubtedly
secured a very flexible medium; for its progressions can be
apprehended either from the point of view of equal steps
and their multiples, or from the point of view of harmonic
trends. The musical effect produced in its use will then
be found to vary according as we have in mind equal intervals or harmonic trends.

"FUTURISM" IN MUSIC

Among modern innovators, the "futurists" in music seem
intent upon a wider employment of the scale, that will lib-
erate us from the traditional restrictions of harmonic principles. A recent manifesto by Balilla Pratella indicates the
nature of this program, which we may quote in the words of
the writer (*90*):

I. It is necessary to conceive melody as a *synthesis of harmony*, in considering the harmonic definitions of *major, minor,
augmented,* and *diminished* as simple details of a unique chromatic atonal mode.

II. We must consider enharmonic music as a magnificent con-
quest of futurism.

III. We must be delivered from the obsession of the rhythm
of the dance, by considering this rhythm as a detail of free
rhythm, just as the rhythm of traditional poetry may be a
model of the strophe in free verse.

IV. By the fusion of harmony and counterpoint we must create absolute polyphony, a thing which never has been tried till now.

V. We must seize all the expressive technical and dynamic values of the orchestra, and consider instrumentation as a sonorous universe of an incessant mobility, constituting a unique whole by means of the real fusion of all its parts.

VI. We must consider musical forms as direct consequences of passionate creative motives.

VII. We must, moreover, be careful not to consider as absolute symphonic forms the traditional schemes of the symphony used to-day, but regard them as decayed and surpassed.

VIII. We must conceive the opera as a symphonic form.

IX. We proclaim as an absolute necessity that the composer must be the author of his dramatic or tragic poem which he sets to music. The symbolic action of the poem ought to issue from the genius of the musician, under the musical impulsion of his soul. A poem written by another would place the composer in the deplorable necessity of receiving from another the rhythm of his own music.

X. We must recognize in free verse the only means of reaching polyrhythmic liberty.

XI. We must translate into music all the new metamorphoses of nature, incessantly and always differently conquered by man in his scientific discoveries. We must express the musical soul of the multitude, of the great industrial factories, of trains, transatlantic steamers, war-vessels, automobiles, airplanes. Finally we must add to the great dominant motifs of the musical poem the glorification of the machine and the victorious kingdom of electricity.

These are the violent and absolute principles that I have eloquently defended behind the footlights of the Italian theatres, confronting the beautiful incendiary applause of our great futuristic adherents.

Despite the bombast and the confusion of this pronuncia-mento, it is a very instructive statement, for it reveals the course that may be taken in certain innovations not alto-gether anarchic. What the futurist demands of his "unique chromatic atonal mode" is only the unrestricted modulations which are possible in the Siamese and Javanese music, and which we, too, can have if we choose to accept the chromatic scale as a scale of equal intervals. Whenever the scale is so employed, "enharmonic" results must follow. "Absolute polyphony" as a "fusion of harmony and counterpoint" means a freer usage of simultaneous tones with reference to the proportional effect of intervals, as well as a less restric-tive employment of harmony; in this way symphonic forms can, of course, be enlarged. Moreover, rhythm is to be de-livered from the "obsession of the dance," "dance" meaning, of course, the simpler measures based on two or three beats and their multiples. With the value of an enlarged dynamic in associating and interpreting "factories, . . . war-vessels, automobiles, airplanes," we are not here concerned. Nor need we subscribe to the writer's complete faith in his atonal mode. Western musical audiences have long been adjusted to harmonics, and we can not suppose that they will sud-denly adapt themselves to any considerable expansion of the musical forms with which they are already familiar. Yet if the principles we have tried to establish are valid, it is not improbable that music will take some such direction in the future. Indeed, everything points to such a development;

for we have only to glance at the history of music to realize
that progress and evolution have constantly been made
through innovations leading to greater flexibility of form
and substance. We here offer no apology for "futurism"
in music: indeed, it is highly doubtful if any considerable
artistic success is ever achieved by revolutionary methods.
Great art is too firmly rooted in all manner of traditions to
find an origin in iconoclasm. Yet the iconoclasts have their
part to play in the destruction of outworn conventions, for
they help to clear our view of the horizon, and they sometimes open new paths of inspiration which the genius of the
future, if not the "futurist" himself, will learn to tread.

MUSICAL IDIOM

Coming back to the essentials of melodic form, we may
say that melody is a succession of tones that interrelate themselves in various ways. The chief laws governing their
orderly progression, or movement, are those of the tonic
trend and the equal interval, supplemented, of course, by
the law of return and the law of cadence. In filling out, in
embellishing, and in affording means of transition from one
key to another, propinquity is an important factor. Together with these influences of tone upon tone, there are also
the volumic, the durative, and the intensive attributes of
sound, that will serve as a basis for musical measure—
metre and rhythm, with their subjective and objective

stresses. By these varied means the most diverse melodic effects can be obtained.

An interesting feature of melodic construction is the employment of conventional idioms. Upon analysis the characteristic effects of folk-music in different lands are attributable to the use of certain scales, or to the employment of certain durative and rhythmical patterns. The typical music of the Scotch is that of the pentatonic scale, a five-interval scale in which semitones are neglected, leaving whole-tone and one-and-one-half-tone steps. The black keys of the piano give us this scale; and melodies composed with these intervals possess a characteristic effect which we associate with the Scotch because their national instrument, the bagpipe, produces these tones.

Scotch air in pentatonic scale

Another characteristic idiom is found in Magyar folk-music. Here the scale consists of the following peculiar intervals: a whole step, a half step, a one and one-half step, a half step, a half step, a one and one-half step, and a half step—in musical notation, c, d, e♭, f♯, g, a♭, b, c. When one compares this scale with the usual diatonic mode, he finds Magyar music distinctive in its employment of more semitones (4 to 2), more tritones (3 to 1), fewer fifths (1 to 4), and fewer whole tones (2 to 5). The use of these intervals

in the peculiar rhythm of the *Czardas* constitutes the basis
of the Magyar idiom.

Passage from Hungarian Rhapsody No. 13.   FRANZ LISZT.

Example of Employment of the Magyar Scale

A third illustration of an idiomatic effect in which a rhythmical pattern seems to be the chief characteristic is well
known to us as the syncopated "rag-time" sometimes attributed to the American negro.

But aside from rhythm there are other elements in the
structure of negro melodies which give a special character to
their idiom. Krehbiel (*51*, *70*) informs us that an analysis
of *527* negro folk-songs indicated seven very frequent variations from the normal or conventional diatonic major and
minor scales. Four apply to the major, and three to the

minor scale. In the major scale, the first is the flatting of the seventh by a semitone. This makes the ascending final interval a whole tone, thus causing the final interval to lose its tonic trend towards the upper tone of the octave. The new seventh is minor in its relation to the key-tone, and also in its relation to the second tone of the scale. The second variation omits the seventh altogether, and the third omits the fourth, while the last major variation is a pentatonic scale having neither the seventh nor the fourth.

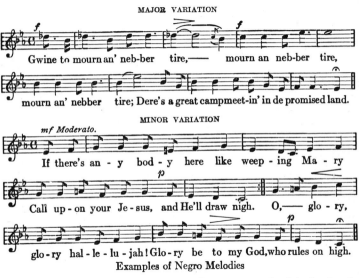

MAJOR VARIATION

Gwine to mourn an' neb-ber tire,—— mourn an neb-ber tire,

mourn an' nebber tire; Dere's a great campmeet-in' in de promised land.

MINOR VARIATION

mf Moderato.

If there's an - y bod - y here like weep - ing Ma - ry

Call up - on your Je - sus, and He'll draw nigh. O,—— glo - ry,

glo - ry hal - le - lu - jah! Glo - ry be to my God, who rules on high.

Examples of Negro Melodies

The general effect of these variations is to abolish the leading tone and the subdominant, thus destroying the tonic effect and establishing an arbitrary final cadence. With the abolition of both the fourth and the seventh we have the usual characteristics of the pentatonic scale, which are mainly atonic in the relationships employed.

Of the variations recorded in the minor scale, the first is a raised, or major, sixth; this substitutes atonic for tonic intervals both with reference to the key-tone and its octave. The second variation abolishes the sixth altogether, and the third substitutes a raised seventh, or leading tone, for the minor seventh, with its reversed tonic relationship to key-tone and octave.

Examples of Idioms of different composers

Aside from the idioms of folk-music, both melodic and rhythmical, such as characterize various national types, we meet also with the employment of distinctive musical pat-

terns by different individual composers; these patterns enable us to distinguish the mode of a Wagner, a Mozart, a
Beethoven, a Verdi, or a Schumann. For one thing, harmonic settings offer many opportunities for individual expression. Furthermore, the interrelations of melodic sequence, harmony, measure, and rhythm are often very complicated, and the idiom is, of course, a function of all three
combined. Although analysis is very difficult at times, it is
never impossible, if we but consider the principles which
underlie musical thought, and the varied ways in which they
operate.

PRINCIPLES OF HARMONY

(a) Counterpoint

Finally we pass to a consideration of harmony, or the
simultaneous employment of tones, where musical thought is
dominated by fusion. Two important polyphonic effects are
here introduced. Of these, one is commonly called *counterpoint*, and the other *harmony*. Counterpoint involves the
simultaneous employment of melodies so set, one over against
the other, that their coincident tones fuse; or if they do
not fuse, they at least give evidence in their dissonance of
a demand for resolution—resolution meaning that certain
tones must follow in order that a tonic or "resting" effect
may ensue. In the case of harmonic accompaniment, tones
coincident with the melody are chosen to support or embellish it. Usually they lie at a lower level of pitch, and

support the melody with fused intervals. While the contrapuntal principle allows the lower tone to assume an equal
or even a greater importance than attaches to the higher
tone of a binary chord, harmonic chords, as a rule, give
prominence to the highest tone, which carries the melody.

Example of polyphonic style—Counterpoint

(b) *Attitude*

With simple bitonal chords the order of fusing intervals
is fairly definite. The octave fuses most completely, then
the fifth, and then the fourth. The thirds and sixths are
more equivocal, their position varying with the attitude of
the hearer. All other intervals are commonly regarded as
dissonant, although for musical purposes the sevenths and
the tritone have a special significance. As pointed out in the
discussion of fusion, familiarity and practice notably increase one's ability to bring tones together into consonance.
But the attitudes we assume towards tonal combinations are
diverse, and the effect of chords varies accordingly. Wilhelm Kemp (*45*) has distinguished by experiment four different effects of bitonal chords, which he calls *fusion, sensory
agreeableness, sensory conformity,* and *harmonic conformity.*

The order in which the intervals fall under these four headings he determined with reference to three degrees of merit —good, fair, and poor. He used chords of the fifth, the fourth, the major and minor third, and the major and minor sixth, and his ratings were as follows: For simple *fusion,* the fifth and fourth were in the first degree, the major third in the second, and the rest in the third. For *sensory agreeableness,* the major third and major sixth were in the first degree, the fifth in the second, and the rest in the third. For *sensory conformity,* the fifth, the major third, and the major sixth were all in the first degree, and the rest all in the third. For *harmonic conformity,* the fifth, the major third, and the major sixth were in the first degree, the minor third in the second, and the fourth and minor sixth in the third, or least harmonious, division. His results are perhaps of more interest as indicating the divergent judgments that may attach to simple intervals when influenced by a particular attitude or point of view, than as a definitive establishment of the degrees of fusion, agreeableness, conformity, and harmony; for as yet we have no knowledge of the specific integration of attributes which must underlie each of these if it is to be accepted as a compulsory perception. The results suggest, however, that several diverse attitudes are possible with respect to the same intervals, and that objective conditions are therefore never a sufficient explanation for any musical effect, however simple it may at first seem to be.

What the objective conditions afford is mere perception

based upon a primary integration of the components of sound which we apprehend as a unit. From this unit we proceed to develop all manner of secondary units with the aid of our varying attitudes, and the associated data of feeling and memory. A multiplicity of distinctions is therefore based upon musical practice and the wealth of our conventional experience.

The work of Pratt, *Some Qualitative Aspects of Bitonal Complexes* (*92*), indicates similar results. Among the various factors contributing to the complexity of the bitonal effect, he distinguishes *smoothness, roughness, simplicity and complexity, pleasantness and unpleasantness, volume, horrisonorousness*—"an intimate integration of auditory quality and ear-kinæsthesis into a closely-knit pattern of vibrant qualitative roughness"—and accessory processes both visual and kinæsthetic.

(c)   Paraphony

According to Pratt, thirds and sixths vary in position more than any other intervals, under the influence of a change of attitude introduced by the instructions given the observer. This result seems to support Watt's view that bitonal chords may be classified as *symphonic, paraphonic,* and *diaphonic* (*133*, 155). The first of these classes includes the octave, the fifth, and to some extent the fourth; the second includes thirds and sixths; and the third all dissonant intervals. According to Watt, symphonic intervals tend so strongly in the direction of unity of impression that, when repeated, they seem to destroy the onward movement of melody. For this reason, successive octaves are unmelodic, and consecutive fifths and fourths are placed under a ban in musical theory. Thirds and sixths, on the contrary, are *neutral* grades of fusion, being neither dissonant nor strongly unitary in their effect. Hence their consecutive appearance does not break the "even flow of analytic concentration" whereby a melody is apprehended.

In Watt's conception of *paraphony* we find a convenient term for the description of the intervals that lie between the symphonic fusions on the one hand, and the diaphonic dissonances on the other. But *paraphony* seems to us more readily understood in terms of melodic trends, such as those of cadence and the tonic, than by exclusive reference to an intrinsic "neutrality" attributable to the fact that these intervals are neither "balanced" like the symphonic nor "unbalanced" like the diaphonic chords.

The ban placed upon consecutive fifths may be at least partially explained by the fact that a new tonality is announced whenever one fifth is followed by another. Such an effect is less pronounced when one major third (4:5) follows another, because the thirds are paraphonic—that is, less unitary than the fifths. Being more readily analyzed, the melodic trend is easily apprehended, whereas with consecutive fifths a break is indicated because each possesses so marked a unity. It should likewise be noted that, as individual intervals, the minor third (5:6) and the major sixth

(3:5) are atonic, while the last paraphony, the minor sixth (5:8), has its tonic above as in the somewhat equivocal fourth (3:4).

Again, the ban which musical usage places upon a fourth from the bass as a bitonal chord involves a similar conflict between the law of cadence, which seeks a tonic in the lower tone, and a new tonic above, which is announced wherever a fourth is employed. Hence the fourth from the bass is again destructive of an "even flow of analytic concentration," because we find it difficult to withstand the influence of a new tonic in conjunction with the fundamental upon which the melody has previously been constructed.

The paraphonies have great musical significance. Their equivocal behavior under the various attitudes which a listener can assume affords rich material for musical composition. Their essential merit rests in the fact that they can be harmonized without destroying the melodic trend. Symphonic chords are too compelling in their unitary effect to be easily subordinated to the melodic flow, while diaphonic chords are characteristically disruptive because their components refuse to harmonize. In employing diaphony immediate resolution is demanded. In the employment of symphony the resolution is already achieved, and the melody brought to a close; but paraphony suggests to the listener a variety of attitudes in accordance with which he is moved now to analyze, now to synthesize. Thus he finds his attitude varying from activity to passivity, from pleasure to displeasure; while he apprehends the sounds of these inter-

vals as either smooth or rough, as simple or complex, as
voluminous in themselves or vibrant with the sensations of
other modalities.

(d)   Concordance

It is generally agreed that the addition of other tones
does not alter the primary integration, or fusion, of a bitonal
pair; yet such an addition does alter the total impression
with its secondary accretions.   The difference between the
major and the minor tritonal chord, for example, is very
marked.   If we follow Stumpf's terminology, and refer to
these as the phenomena of *concordance* and *discordance,* we
may note certain uniformities that have been demonstrated
by experiment.   Concordance has been found to be less
when the poorer binary fusion is the lower, and greater when
it is the higher of the two combining intervals.   For example,
the major chord, c-e-g is more concordant than the minor
chord c-e^b-g.   Concordance is likewise increased when the
greater frequency-ratio occupies the lower position, and decreased when it occupies the higher position in the chord.
Thus, c-f-a, whose ratio-numbers may be given as 12:16:20,
is more concordant than c-e-a, whose ratio-numbers are
12:15:20.   The same is the case when we reckon frequency-
difference rather than frequency-*ratio,* as has been shown
by the experiments of T. H. Pear (*87*).   Furthermore, Max
F. Meyer (*67*) found that the concordance of three clangs
will be greater, the simpler the ratios of the frequencies of

vibration, whether the chord be considered as a whole, or its tones taken in pairs—for instance, the common triad c-e-g, $(2:3:4)$ is more concordant than the chord of the six-four c-f-a $(3:4:5)$.

(e) *Summary*

These are only a few of the many uniformities that prevail in the expression of musical thought. All of them derive from the laws of tonality, yet each suggests the manifold diversity of melodic, polyphonic, and harmonic developments which enter into the unique process of musical thought. In the elaboration of principles of harmony, use is made of our natural propensity to expression; and through expression a means of communication is found, because other persons of like constitution are ever ready to receive and to assimilate sounds uttered or otherwise generated in sequences and with accompaniments suited to their natural ways of hearing.

Education, to be sure, plays a large part in increasing one's ability to think, express, and understand sounds; and when we have laid the foundations of musical thought, we have but given the bare outlines of its logic. Not only are the elaborations of usage variously conditioned; they involve many arbitrary signs to designate conventional meanings; and they include, also, a great many compromises. In accepting the tempered scale as a substitute for the scale of *just* intonation we have made such a compromise in order to accomplish something which could not be so readily accomplished with truer intervals; but we have, of course, denied ourselves something in effecting this compromise.

As an illustration of an equivocal result arising from such a compromise, consider the chord c-e-g♯. This chord has the range of an augmented fifth which in tempered intonation is identical with a minor sixth, and it embraces two major thirds. Since all three intervals are individually harmonic, one might naturally expect concordance; yet the effect is a discord. Only by reference to the principles of harmony do we detect the discrepancies which here operate to render concordance impossible; for analysis will show that the augmented fifth has the ratio of 5:8, and this interval can not possibly be divided into two equal intervals of a major third, or 4:5 each. Taken separately, each of the binary pairs in tempered intonation represents the ratio of the third; but when the two are combined a discrepancy is felt, because the upper tone can not at once serve as a major third above the middle tone, and also as an augmented fifth above the lower. In order to make this quite clear, we may refer to the intervals in terms of "cents." In tempered intonation e represents an interval of 400 cents above c, while g♯ represents an interval of 800 cents. In *just* intonation the c should have 386 cents, while the g♯, as a major third above e, should have twice this number, or 772 cents. But g♯ must at the same time mark the interval of a minor sixth from c, which interval has 814 cents. Hence the tone g♯ of 800 cents must serve at once for a flatted note of 772 cents, and for a sharped note of 814 cents. It can do either separately, but it can not do both at once—the discrepancy is too great, the ear rebels, and the combination is pronounced discordant. If it be objected that this chord nevertheless is fairly effective, we must refer its effect to the principle of equal intervals. The two equal steps which comprise the larger interval may explain our interest in the chord, but in strict harmony it still remains a discord. Where harmony is involved, we require a close approximation to the harmonic intervals, but where only a melodic sequence is in order, we not only accept the compromise of equal intervals with readiness, but can even derive a certain satisfaction from our ability to recognize that they are equal.

Music, then, is but the elaboration of certain principles, all of which are founded in sense-perception. Because of its peculiar derivation from the act of perception itself, rather than from the objective or common-sense meanings of external situations, music is regarded as the most abstract of all the arts. Yet it is abstract only when we regard it from the point of view of practical affairs and ordinary adjustments. Tonal experience is not in itself an abstraction. As John Dewey has said, abstractness and concreteness are psychological affairs, the concrete being "whatever appeals to the mind as a whole, as a self-sufficient centre of interest and attention." Surely, music makes its appeal concretely enough to those who understand it, and if it remains but an abstraction to many, this is because they can not or will not devote to it the necessary attention to make it a concrete experience. Consequently, if music exists at all in the minds of such persons, the effort required for its apprehension is too great to yield them any considerable satisfaction. These are they who demand music with some accompaniment of poetry, dramatic action, or the dance; and popular descriptive music, in which the sounds of nature are imitated, is written largely for their delectation.

PROGRAM-MUSIC

Program-music and "tone-painting" have a place in the art of music, but as Busoni (*19*, 12) writes: "These means of expression are few and trivial, covering but a very small section of musical art. Begin with the most self-evident of all, the debasement of tone to noise in imitating the sounds of nature—the rolling of thunder, the roar of forests, the cries of animals; then take those somewhat less evident, symbolic imitations of visual impression, like the lightning-flash, springing movement, the flight of birds; again, those intelligible only through the mediation of the reflective brain, such as the trumpet-call as a warlike symbol, the shawm [or oboe] to betoken ruralism, march-rhythm to signify measured strides, the chorale as a vehicle of religious feeling. Add to these the characterization of nationalities—national instruments and airs, and we have a complete inventory of the arsenal of program-music. Movement and response, minor and major, high and low, in their customary significance, round out the list. These are auxiliaries, of which good use can be made upon a broad canvas, but which, taken by themselves, are no more music than wax figures may pass for monuments."

The investigation of H. P. Weld (*137*), in which various observers were requested to state their impressions upon hearing a piece of descriptive music, revealed an absence of anything like a definite idea of what was being described. But whereas the language of music is incapable of contributing ideas of persons and things, it does not follow that such ideas may not be suitably clothed in a musical garment which adds its own peculiar charm and enrichment to the experience. Without going to the extreme of maintaining, as did the Viennese critic, Eduard Hanslick (*29*), that the only music worthy of consideration is *absolute* music—a music freed from all associative connection with the rest of life—we may allow that music is the language of emotion, and that it provides a melodic, a rhythmical, and a harmonic setting for the dance and the drama. Yet, over and above all these, it exerts an influence in and of itself; for music is a true and independent type, able in its own right to quicken the mind, and to afford a delightful exercise of thought. As Combarieu says, music is thinking in tones; but thought is an extensive and complex structure. While we think in tones, we can also be thinking in words and deeds. In order that unity and coherence may obtain for the whole, some of these trends of thought must be subordinated to others which for the time being are the more insistent and commanding. Although music alone is an inappropriate mode of conveying representative ideas, the distinctive ideas which it creates and conveys are not less significant and expressive in their own way than are the words and actions with which we are wont to conduct our practical life. Musical thought differs from other modes of thought less in the nature of its process than in its character of reference; for music primarily refers to the tones themselves, among which the relations of musical thought are established, whereas words refer also to the objects of practical life which they symbolize.

CHAPTER XI

THE LOCALIZATION OF SOUND

MONAURAL SOUND

THUS far no direct reference has been made to the localization of sound. Although sounds are voluminous in their nature, their phenomenal structure provides no adequate basis for judgments of a real extension comparable to that of visual phenomena. The reason is that sounds have no definite contours; their outlines are even more blurred and vague than the outlines of a shape pressed against the skin; and what is commonly known as "local sign" in the visual and tactual fields seems to be lacking, or at least ill-defined, in the field of monaural sound. Yet, just as the two eyes co-operate to provide a means of judging depth, so the two ears co-operate in supplying data for a judgment of direction. With one ear alone, judgments of direction are much less accurate; still they can be made. We shall therefore devote our attention first to monaural localization before we proceed to the phenomena of binaural hearing.

Evidently the attributes of a phenomenal sound will vary according as its physical source is near or remote; distance renders sounds faint, while nearness makes them louder. The alteration the sound undergoes as it approaches or moves away from us is primarily a matter of intensity; but with variations of intensity there are also bound up variations of all the other attributes. When intensity decreases, the sound appears to diminish also in volume and duration, thus shifting towards a higher level of pitch-brightness. It has often been observed that faint sounds seem to be higher than louder sounds of the same vibrational frequency. This phenomenon may perhaps be attributed to a modification of the wave-length, if such a thing can happen when physical intensity is decreased. That would account for an increase of brightness and a decrease of volume. But we know, as yet, too little of the modifications of wave-form which may be conditioned by a variation of intensity to warrant more than a guess at an explanation. Intensity, on the other hand, seems to contribute to the volume or fullness of a sound, giving it a lower pitch-brightness than it would possess were the intensity less. When tones are of a uniformly high pitch, however, alteration of intensity does not appreciably affect their mass. So discrimination on this basis is lacking in the higher regions of the scale.

Together with these variations in the attributes of sound, a change in timbre is also noticeable when like sounds are heard at different distances. As a sound grows weaker, its partials are also weakened; therefore it becomes less resonant, and may lose some of its original components altogether; and this might be expected to *lower* its pitch— though the effect is of an opposite sort, as has just been stated.

To phenomena such as these we owe our ability to estimate the direction and distance of sounds heard with but one ear. In the case of direction, the judgment is largely based upon the impression the sound makes when the head is turned; for a sound is heard to best advantage by one ear when the auricle, or *pinna,* is directed toward the source of the sound. In some animals movement of the auricles is possible, and they will prick up their ears, so that the direction from which the sound emanates can be detected without moving the head; but man's auricle is degenerate with respect to movement. The muscles are there, and can be trained to produce slight movements, but these movements are quite insufficient to affect the sound noticeably. Indeed, one's hearing need not be seriously impaired if the auricle is entirely removed; though the auricle does serve to some extent in deflecting the sound inwards, so that sounds coming from the front are somewhat more distinct than sounds coming from the rear. In other words, we possess a certain capacity for differentiating sounds that come from in front from similar sounds coming from behind; and this capacity prompts us to turn the head until the sound observed reaches its optimal fullness of intensity and resonance when its direction coincides with the aural axis of one ear. The direction of the source is then inferred from the position of the ear. It should be noted that a certain pattern of resonance is often effective even when no change in the attributes of a sound can be observed to accompany movements of the head. The timbre of a tone, or the corresponding mixture of sound-ingredients in a noise, characterizes the impression in a way that indicates its meaning—as being that of a whistle, a bell, or a shout—and will often materially aid in determining both the distance and the direction of the sound; for resonance is very suggestive and from it appropriate inferences may be drawn. The sound of a whistle may or may not be clearer and fuller when the head is turned towards its source; but if the sound is like a locomotive-whistle, an estimate of its direction and distance is readily made if we are able to assign it to the already known location of a railroad.

BINAURAL SOUND

With binaural hearing additional data for our judgments are supplied by differences both of attributes and of timbre, as the sound is heard by the two ears. These binaural effects are termed *diotic* when the two ears function together in the same manner—that is, when the same sound is heard in each ear. They are termed *dichotic,* however, when a perceptible difference in the phenomena can be made out—as when the two ears hear different sounds simultaneously. When the position of the head is changed, alterations are produced in the phases of the partial vibrations as they strike the two ears, and the total effect of the clang is thereby modified. An appropriate interpretation of this modification will enable us to detect the direction from which the sound issues. With some sounds, however, one can not discern any differences, even though the sound itself is distinctly characterized. Take the chirp of a cricket, for instance, or the cry of a katydid. On a quiet evening the air may seem filled with such a sound, and yet we are quite at a loss to locate it, the reason being that the sound is of very high pitch, and is also very bright, shrill, and piercing; consequently, a rotation of the head produces no change in the sound's attributes, while differences of timbre are likewise lacking, because the overtones, being still higher than the fundamental, have no audible effect.

The author once had an experience which well illustrates this point. Awakening one still summer night, I noticed a faint but persistent high tone ringing in my ears. As I turned from side to side on the pillow, the sound continued unaltered. My first reaction was only that of a slight annoyance at being disturbed, but gradually I became apprehensive, lest the cause should be a *tinnitus* of more or less serious import. Unable to sleep, I arose and went about the house, listening from window to window, at the front, at the back, and at the sides of the house. All the windows were open, and at each the obsession of this high-pitched continuous sound persisted. Finally, overcome with weariness, I returned to bed and fell asleep, with the depressing conviction that some auditory derangement had befallen me. In the morning the sound had gone, and was thought of no more until later in the day when one of my neighbors casually mentioned that he too had been similarly disturbed by a sound which he had traced to an exhaust of steam that had been allowed to pass through a whistle-valve in a somewhat remote manufacturing-plant. The sound was so high in pitch that it lacked audible overtones, and yet so penetrating that in observing it from different sides of the house its intensity suffered no appreciable impairment. For this reason quite naturally I placed it within my own head, this being the usual judgment as to the location of sounds which do not vary with a change of bodily position.

Ordinarily, binaural co-operation aids us in localization whenever the sound is not confined to the median plane of the head, between the ears. The direction of a sound is therefore frequently determined with some degree of accuracy by moving the head until the sound is heard equally well in both ears. We judge it then to be in the median plane, and usually in front. The difference between *right-side* and *left-side* sounds is the most striking feature of binaural localization, and this suggests the question as to the precise nature of binaural co-operation.

Von Liebermann and Révész (57) have found that with tones in the neighborhood of 1,000 v.d. most persons of normal hearing report a difference, often amounting to a semitone, when a sound is heard alternately in either ear. If the ears function together, the resultant sound has the combined intensities of the two separately heard tones, and assumes a pitch midway between them. From this fact these authors argue that the mixture is analogous to the mixture of visual impressions in the two eyes, and that there may be "corresponding points" in the field of hearing, like the "corresponding points" of visual space-perception.

That is to say, a discrepancy in the pitch of two monaural sounds is supposed to furnish the true basis of our judgment as to the distance of its source from us. We may doubt if such an explanation is needed, when we consider how slight is our ability to detect the depth of sound. Baley (6, 7) obtained experimental results similar to those of von Liebermann and Révész, but was inclined to believe that the binaural mixture could be explained by a spread of resonance about a maximal point, heard in each ear. Using clearly distinguishable musical intervals, he found that trained observers could distinguish as many as ten simultaneous tones, and refer them correctly to the appropriate receptor, when but a part of them had been conducted exclusively to one ear, and the rest to the other; which would seem to indicate that the mixture of binaural sounds is more apparent than real. Yet differences in the binaural impression must furnish whatever data we can make use of in determining the distance and direction of a sound more accurately than we can with one ear alone—just as differences in the binocular impression must furnish the data for our judgments of visual space. It is doubtful, however, if a dichotic difference of pitch is any more important than dichotic differences among the other attributes.

When a single tone is heard alternately in each ear, a modification in its intensity may also be detected; usually one ear hears it louder than the other, as well as higher or lower in pitch; and Otto Klemm (46, 47) has shown by experiment that the intensity of the diotic sound, heard simultaneously in both ears, is more than twice as loud as the same sound when heard by one ear alone. Klemm's experiments also show that when a sound enters one ear about five-thousandths of a second in advance of its entrance into the other ear, its direction is indicated by reference to the receptor which first hears it. Thus the attributes of intensity, pitch, and duration all furnish data whereby the direction of sound can be determined.

In addition to these, the attribute of volume has received special consideration from Watt (*132*), who has reached the conclusion that it is, indeed, the most important factor of all in the discrimination of sounds belonging to the right and the left ear, respectively. According to Watt's theory the volume of the higher sound finds its place *longitudinally* within the volume of the lower sound, or, if there chance to be identical sounds, these will coincide, just as in a monaural mixture. But the fact that even identical sounds are somehow heard differently in the two ears suggests to him that in their *transverse* aspect the volumes of two monaural sounds overlap without coinciding. This, he thinks, makes it possible for us to detect a difference between the right-side and the left-side impressions of a single objective sound, which could not otherwise be so distinguished.

THE PHYSICAL CONDITIONS OF BINAURAL SOUND

Let us now consider the physical conditions under which a binaural sound is localized. Primarily these are three in number: (1) the intensity-ratio of the sound at the two ears; (2) the phase-difference of the sound-wave at the two ears; and (3) the relative time of arrival of the sound at the two ears. The earlier investigations of physicists tended to support the hypothesis that the intensity-ratio was the most important factor in the localization of sound, but this hypothesis is now being abandoned. In more recent experiments, greater attention has been paid to the purity of the tone employed as the source of sound. Means have likewise been devised to secure a variation of intensity without a corresponding variation in either phase or time. According to the observations reported by Halverson (28), judgments of direction are continuous and fairly accurate in the region of the median plane, but the sound appears to reach a fixed location within 30° of this plane, where it remains until the intensity of one sound is so overwhelming that a second sound is apprehended in the aural axis of the ear where the intensity is greatest. A sudden perception of movement may accompany the introduction of the second sound, but the first sound nevertheless remains in its fixed position until its faintness, with reference to the second sound, causes it to become inaudible. Thus the localization of the sound, as conditioned by a gradual variation of the intensity-ratio, is essentially discontinuous. The sound first spreads in the direction of the greater intensity, but soon comes to a halt at a position not far from the median plane; and upon further increase of intensity, a second image appears in the aural axis. By a gradual decrease in the dichotic ratio, a reversal of this effect is produced. Again, the sound jumps as a second image appears near the median plane, and thereafter the image in the aural axis fades away, leaving only a single sound in front of the observer.

Quite different is the effect when the phase of a tone is altered without change of intensity. Here localization follows the advancing phase. For instance, if a single "pure" tone is conducted by means of tubes to each ear, and a device is introduced for lengthening or shortening the tubes so that the phase of the tone in one ear may be advanced over that in the other, the sound thus heard is localized in the median plane when there is no difference of phase. As the phase advances in one ear, the sound is localized as moving in the direction of the aural axis of that ear. Lateral localizations are more difficult than frontal, and the "phantom image," as it has been called, "appears to be auditorily more intense, voluminous, and diffuse than when frontally perceived" (*ibid.*, 187). "When the phase-difference at the ears approaches 180° (opposition of phase), the tonal image attains its most lateral position, and gradually disappears. It is succeeded by a second image at the other side of the head, when the difference in phase is again less than 180°, and the phase at the latter side leading. In the intermediate critical position, double images, one on either side of the head, may be observed. When phase alters rapidly through this critical position, the rapid succession of images may produce an illusion of movement of a single image through the head or behind it" (*ibid.*, 187-8).

Judgments of localization with reference to phase-difference are found to be much more accurate than those depending upon the intensity-ratio; and Stewart (*115*) states that, for tones lower than 1,200 v.d. to 1,500 v.d. (above which level the sound-image is confused), "the angular displacement from the median plane produced by a given phase-difference at the two ears proves to be the very position for the source of sound that will theoretically produce the aforesaid phase-difference at the ears" (*ibid.*, 441).

Stewart believes also that the effect of phase can not be that of the time-interval involved, because equal time-intervals do not correspond with equal angular displacements, except at higher frequencies than those employed in his tests. He therefore concludes that the localization of sound is chiefly dependent upon a direct perception of phase.

The influence of the time-interval has been recently investigated by von Hornbostel and Wertheimer (*41*), who employed a similar method with a noise instead of a tone. The noise in question was a short sharp clap like Abraham's "simple" noise, which was found to be conditioned by a modification of wave-length. With time-intervals varying from 0 to $630\sigma\sigma$ the noise gradually shifted from the median plane to the aural axis of the ear which first received it. The threshold of change was approximately $30\sigma\sigma$,[1] and when one monotic sound was $63c\sigma\sigma$ in advance of the

---

[1] 30 $\sigma\sigma = 30 \times 10^{-6}$ seconds, and is the time-difference between two sound-waves in the air when one of them is 1 cm. in advance of the other.

other, the binaural effect was localized in the aural axis of whichever ear received the sound first. Alterations of intensity in the dichotic sounds had no marked effect upon the judgment. When the advanced temporal stimulus was weakened, localization was judged as before, until the sound became inaudible; then the location shifted to the other ear in which the retarded noise was still to be heard. With tones, phase-differences must accompany time-intervals, but tones were found to be less accurately localized than the simple noise, and the authors conclude that the time-interval is a better criterion than phase, even with tones.

THE THEORY OF BINAURAL LOCALIZATION

We shall attempt no reconciliation of this view with the opposite one advanced by Stewart; but the remarkable uniformities reported with reference to the time-intervals of a simple noise suggest that the common feature of noise and tone, the wave-length, may prove to be the critical condition for the phenomenological variations upon which the localization of sound depends. Both phase- and time-difference at the two ears could affect the dichotic pattern, and if further investigations indicate that the binaural integration of sound involves a phenomenal variation in *brightness* when the time- and phase-relations of the stimuli are altered, we shall then have an attributive criterion for our judgments of direction and location.

Whether the impression of direction is a product of binaural receptivity, or of cultural organization, is an open question; but some phenomenological criterion there must be, and it must somehow be registered in the sound heard. For this reason the psychologist can not be content with a uniformity of judgment which is shown to vary directly with either time-interval or phase-difference. Nor is it enough to speak of a "direct perception of phase," or a "direct perception of time." To the psychologist who is concerned with an integration of attributes which will supply a phenomenological basis adequate to his perceptions these are but statements of problems. *Phase* is not a psychological phenomenon, nor has any attribute of sound been discovered which varies directly with phase.

Nor is *time* psychological, though *duration* is. But the time-interval, in accordance with which sound is localized, is far too brief to be perceived as a duration. What concerns us, therefore, is the influence exerted by phase, or time, or both, upon the dichotic sound emanating from a source, the direction of which is indicated by these physical variants. A perception of phase or a perception of time in a binaural impression is, therefore, a complicated integration of the attributes—pitch, brightness, duration, intensity, and volume—in which one or more of these attributes will vary directly with phase or with time to alter the phenomenon of sound in a discriminable manner. We can only guess what this alteration involves; but if pitch, duration, and intensity are perceptibly constant in a tone that is nevertheless perceived as having different locations when its time or its phase is different at the two ears, we may infer that the sound has been somehow affected with reference to its brightness or volume, and perhaps both. Since the vibration-frequency is not altered at the two ears, volume, from what we now know of its conditions, is less likely to be the attributive variant than brightness, which seems to be dependent on wave-length. An integration of the attributes of a sound occasioned by waves at the two ears, one of which is slightly advanced in phase or in time over the other, might well produce the modification of the "primary wave" which Abraham describes. This might occasion a reduction of brightness in the sound, and thus provide a phenomenal basis for judging the trend of this reduction in the one ear as compared with the other—a conclusion that seems to be in some measure indicated by the "diffuseness" of the lateral sound as reported by Halverson's observers. But subjective changes in the direction of greater intensity and greater voluminousness, also reported by his observers as characteristic of the lateral sound, may likewise be correlated with physiological and psychological modifications in the sound-picture. Yet of these we have at present no data upon which to theorize.

We must therefore be content with an analysis which indicates that time-differences are very important conditions for binaural localization; that when the sound is a tone, phase-difference is also involved; and that the intensity-ratio may likewise be a factor. On the psychological side, all the attributes of sound are involved in the binaural pattern, and, separately or in conjunction, each may serve as a means of detecting the direction from which the sound emanates. In considering that these attributive differences involve all the components of a clang, both its partial tones and its noisy constituents, we can realize that the data are fairly numerous upon which judgments of localization rest. Our inability to localize sounds with accuracy is therefore not one of incomplete data, but one of faulty discrimination. Since the localization of pure tones is found to be much more difficult than that of clangs and noises, it follows that the complexity of the constituents is itself an important consideration; for in complex sounds there are several pitch-salients with their respective intensities and temporal aspects, as well as peculiarities arising from the resultant pattern of volume, all of which afford diverse means of discrimination. In judging the localization of a sound, we consider the impression as a whole, rather than some particular feature or features that emerge clearly into consciousness; we do not consciously analyze the impression; we accept it as it is, and draw our inferences intuitively. At best, the ear does not localize well; and whenever we reach an assured judgment of distance and direction, we are apt to be drawing largely upon information otherwise secured than by the direct means of hearing. As I write, I hear a sizzling sound at my side, but its localization is not altogether determined by the more striking sound-picture in my left ear; for the sound at once indicates an escape of steam, from which I infer its location at the radiator, the position of which in the room is well-known to me.

In the realm of the familiar and easily authenticated facts of everyday life, our judgment of localization is practically effective, though never very accurate. When we simplify the conditions by a laboratory experiment, we begin to understand for the first time what these variable conditions are under which our intuitions of direction and distance are made. Thus we find intensity less effective than time, although the two may function together in a practical situation along with many other accessory circumstances. When his task is to locate a sound in a natural environment, the intuitions of a keen observer may be far more accurate than any that are made under the greatly simplified conditions of an experimental observation. Yet these intuitions, being made upon the basis of unanalyzed data, are likely to be much less uniform with a changing environment than are the carefully analyzed and discretely directed judgments of the experimental observer whose environmental conditions have been controlled.

We see, then, in the localization of sound a very pretty example of the complex integrative results arising from very minute variations of physical stimulation. Yet while no complete point-to-point correspondence can now be established between the stimuli, the physiological processes, and the psychological phenomena of sound, we can trace the variations of each attributive component to a degree which reassures us in our belief that there is a correspondence, and that the psychological phenomena underlying the localization of sound consist of data which interact with one another quite as definitely as do the physical data of the sound-waves themselves.

CHAPTER XII

THE PATHOLOGY OF HEARING

THE DIAGNOSIS OF AUDITORY DEFECTS

DEAFNESS, a common ailment, is one that frequently resists medical treatment. It may be either congenital or acquired, and its special occasion either a structural defect or a functional derangement. The defect or derangement may involve the sense-organ, the nervous connections between sense-organ and brain, or both. In this brief summary we shall not attempt an exhaustive analysis of the pathology of hearing, but confine ourselves to typical defects and their diagnosis.

As already noted in Chapter II, the organ of hearing may be roughly divided into three parts: (1) the external ear, consisting of the auricle and the external auditory canal, or meatus; (2) the middle ear, or tympanum, with the drum-membrane and ossicles; (3) the inner ear, containing the cochlea and the vestibular apparatus. A defective mechanism, or a functional derangement of any of these parts, may occasion deafness or some other anomaly of hearing. But defects are likewise attributable to the nervous system, both along the afferent tracts to the brain, and in the collateral tracts which function in auditory perception. A general distinction, however, may be drawn between defects of the auditory mechanism external to the cochlea, and those from the cochlea inwards which involve the auditory nerves.

In order to determine whether a case of deafness has its origin in the conducting apparatus or in the nerves, certain diagnostic tests are made. One of these is the so-called *Weber* test. A vibrating tuning-fork is placed in the sagittal line on the vertex of the skull, and the patient is called upon to decide in which ear he can best hear the tone. Since the sound is in part conducted through the bones of the skull, it will be heard in each ear unless there is a cochlear or central nervous defect, but it may be heard better in one ear than in the other. If one of the external canals is closed, the intensity of the sound in the closed ear is increased by reason of the fact that the passage of the sound-waves outward is to some extent interfered with, so that the imprisoned column of air resonates, and a second reflection of the sound may ensue towards the labyrinth. A similar effect may be obtained when there is a stoppage of the auditory canals; the sound of the fork is better heard in the affected ear than in the unaffected organ, from which the sound passes freely outwards. If the disease has its seat in the inner ear, however, or along the central tracts of the auditory nerves, the sound will be heard best in the unaffected ear.

Another test, associated with the name of *Schwabach*, consists in a comparison of the duration of a sound by bone-conduction in the diseased ear and in a normal ear. The fork is again placed on the vertex of the skull, and defects are then indicated either by marked prolongation or by marked shortening in the duration of the sound heard by the affected organ. When a prolonged sound is heard, the test is first made with a normal hearer. After the sound has ceased for him, the fork is transferred to the skull of the patient, and the additional duration of the sound is measured so long as the patient continues to hear it. When the defect indicates a shortened duration, the fork is first placed on the skull of the patient, and then transferred to the normal hearer, the additional duration again being measured. Thus a comparison is established between normal and abnormal hearing under like conditions. In practice, this test is simplified by applying a fork whose duration is known to the diagnostician from tests previously made on the normal ears of many persons of different ages.

A third test, that of *Rinne,* consists in a comparison of the duration of sound by bone- and by air-conduction, respectively. The normal ear being more sensitive to air-conduction than to bone-conduction, after the sound has ceased to be heard with the fork resting on the skull, it may be heard again for thirty seconds or longer if the prongs are brought before the external auditory meatus. The superiority of air-conduction over bone-conduction becomes more marked the lower the pitch of the tone employed. In order to make the transfer readily, the sound is measured with the fork resting on the mastoid process rather than upon the vertex of the skull.

This test gives the most refined indication of derangements in the conducting-apparatus; for every change of equilibrium in the auditory mechanism prolongs hearing by bone-conduction, and shortens hearing by air-conduction. If, however, the defect is in the inner ear, the difference between the prolongation of sound by bone- and by air-conduction remains virtually the same, both being reduced in about equal degree. As deafness increases, on account of an internal defect, hearing by bone-conduction finally ceases altogether, though in some cases the sound is observed to persist after it can no longer be heard at all by air-conduction.

By means of these empirical tests it is possible to determine, roughly at least, which of the two ears is affected, or, if both are deaf, which is the more so. One can also infer whether the defect is attributable to the sound-conducting mechanism, or to the inner ear and its nervous connections.

It is not definitely known whether sounds conducted by the skull have an immediate effect upon the cochlea, or if, like any other sounds, they are mediated through the mechanism of the middle ear. Bezold (*13, 65*) remarks that bone-conduction is probably but an indirect means of communication through the usual channels of the middle ear, differing from air-conduction only in that the sound-waves strike the *edge* of the drum-membrane and the *ligamentum annulare,* and not the flat surface thereof. Bing (*14*) has reported a clinical case in which the drum and ossicles of one ear were entirely destroyed, yet the patient could still hear with that ear, presumably by bone-conduction. This case, however, can be reconciled with Bezold's views if we assume that the vibration of the skull was sufficient to set up corresponding air-waves in the middle ear, which in turn were communicated through the oval or the round window, or through the *promontorium,* to the cochlea. It should be borne in mind, of course, that while certain portions of the skull are dense, and therefore good resonators, other portions, being porous, lack this physical property. What is commonly known as bone-conduction involves the placing of a vibrating instrument on the skull, a more complex method of stimulation than the ordinary communication of a sound-wave propagated through the air and communicated through the canals of the ear (*143*).

ACUITY OF HEARING

In order to measure the acuity of hearing in one ear or both, tests are made with the chief types of sound—tones, vocables, and noises. Tones are principally used to determine the range of hearing, and to discover such anomalies as tonal gaps. The sensitivity to tones in the higher range of pitch gradually decreases with age. Thus Gildemeister (*26*) found that children could hear tones of 20,000 v.d., but persons in the middle thirties rarely heard a tone above 15,000 v.d.; while at the age of fifty the upper limit had been reduced to about 13,000 v.d. By bone-conduction the hearing of tones was always found to be a few hundred vibrations lower than by air-conduction. These results were obtained for a constant medium intensity; with increased intensity, however, audible pitch could be indefinitely raised.

The appearance of tonal gaps and tonal islands within the range of hearing is an interesting anomaly which a careful examination of the whole range of tonal hearing sometimes reveals. Bezold (*13,* 53 ff.) cites twenty-five cases which he found among fifty-nine deaf-mutes. In eighteen of these, the examination indicated a single patch, or island, of hearing extending over an interval that varied all the way from a musical fourth to nearly three octaves. The location of these patches varied from the main octave (70 v.d.) to the fifth prime octave (5,000 v.d.). Bezold also found six cases possessing two islands with a gap between, and one case with three islands. It is of special interest to note that these anomalies show no marked tendency of location in any specific region of the total scale. The islands of different cases overlap, and when taken together extend throughout the musical range of sound.

The measurements of acuity to noise are undertaken in a rough-and-ready way with the aid of a watch held at different distances from the ear. More exact instruments known as *audiometers* enable us to standardize these tests with greater precision. Seashore (*106,* 87, 90) has perfected such an instrument, and also one in which the pitch-range may be tested with a variable intensity for each pitch.

Since deafness is particularly significant with respect to the sounds of the human voice, tests of hearing speech-sounds are of especial interest. As we have seen, the sounds of speech are of a peculiar complexity, and involve both a range of pitch and a nice balance of intensities at the different points in the scale of hearing which mark the regions of the vocal formants. With the precise knowledge concerning these formants which is rapidly becoming available, it should be possible to make a complete and satisfactory examination with the aid of such an instrument as Seashore's pitch-range audiometer. Lacking this special knowledge in the past, diagnosticians have usually relied either upon the whispered word, or upon artificial vocal sounds as produced, for instance, by the vowel-siren of Marage (*62*).

In his analyses of speech, Stumpf (*124*) has measured the limits of comprehension by a gradual reduction of the range of pitch within which words can be heard with understanding. By eliminating the higher-pitched sounds, he found that a complete understanding of words is possible when no sounds heard are higher than $d^{\#4}$, at about 2,461 v.d., although at this level the individual vowels *i* and *ü* have become *u*-like, and *e, o*-like. A further reduction to $g^{\#3}$ (1,642 v.d.) gave a hazy effect to speech, yet with concentrated attention everything could still be understood. The most decided change was noticeable when eliminations occurred in the upper part of the second prime octave, in the region of the formant for *a*. This region, which he terms the "speech-sixth," e^2-c^3 (cf. Fig. 13), appeared to be the most important of all, for with its destruction the capacity to understand the spoken word was lost. Yet it was remarkable how long speech could still be understood even after a disturbance of this region had set in. Apparently, the struc-

tural conditions of the vocal phenomena are so ingrained
that just as "a run helps a jump," so a dispositional readiness
helps us to complete a very defective sound-pattern if only
the words occur in some familiar sequence.

Bezold (13, 302) had previously discovered in his practice the importance of a similar region of sound, but the
speech-sixth as determined by him (b^1-g^2) is about a half-
octave lower than Stumpf's. The greater exactness of
Stumpf's method of measurement, however, enables us to
regard his interval as the more trustworthy.

According to Bezold (*ibid.*, 72), all whispered words can
be perceived by the normal ear in a comparatively quiet
room remote from the noises of the street, at a distance of
20 to 25 metres. Examining 1,918 school-children—3,836
organs—by this method, he found that 26 per cent possessed
only about one-third normal hearing.

Since the sounds involved in the spoken word are so complex, verbal perception may be still possible, as Stumpf
found, even when a marked defect of hearing is indicated,
provided that the patient has learned to make appropriate
use of whatever sounds he does hear. In this respect sur-
prising results are claimed by French otologists (*56, 61, 63,
94, 95*) in the way of re-education effected with the aid of
the vowel-siren. With daily practice of not too long duration
in hearing the chief vowel-sounds, the patient is said to show
a marked and rapid improvement in his capacity to perceive
words. Beneficial results from this training are claimed
for all sorts of deafness, including both structural and func-

tional defects. Certain deaf-mutes indicate by their response to the treatment that their failure to speak and to understand the spoken word is attributable to improper guidance in perceiving those sounds that have a peculiar significance in oral communication. Marage (*61*) reports that 68 per cent of a group of unselected cases of deafness, incident to the late war, were cured by this treatment, and enabled to return to the front. Likewise Ranjard (*95*) reports 84 positive results from 100 cases of soldier's deafness. According to Zwaardemaker's experience (*150*), the use of the whispered word is even more efficacious in such treatment than the vowel-siren.

THE NATURE OF AUDITORY DEFECTS

These being the chief methods of diagnosis, we may now proceed to consider the nature of certain common auditory defects. Deafness arising from a defect of the sound-conducting mechanism may have its seat in the external auditory meatus, in the drum-membrane, in the ossicles, or in the middle-ear cavity. Stoppage of the canal, and defects of other parts of the apparatus, may be congenital; more often they are a result of suppurations and inflammations caused by micro-organisms introduced through the auditory passages or through the blood. With children the introduction of water into the outer canal of the ear is a frequent cause of sore ears, attended by a formation of pus, which may interfere with hearing. In bathing, it is inadvisable to allow any water to enter the canal, nor should oils and other fatty substances be introduced except with the advice of a physician; both moisture and fat being conducive to the propagation of micro-organisms which cause a disease of the tissues that may extend inwards through the drum-membrane into the middle ear, or even into the inner ear.

The numerous diseases of the drum-membrane may be brought about by mechanical injury, by piercing or bruising the membrane with a sharp instrument, or by a rupture of the membrane, caused by a sudden and intensive change of air-pressure upon it. A blow on the head, a dive into the water, or the detonation of an explosion frequently causes a rupture of the drum. The healing of the tear or puncture is commonly effected in a short time without adverse effects, provided that no infection has taken place in the tympanic cavity.

According to Bezold (*ibid.*, 130), 66 per cent of all his patients suffering from ear-troubles indicated some affection of the middle ear. The sensitivity of its membranous tissue renders it peculiarly susceptible to infections, especially in childhood. *Otitis media* may be *simplex* and non-suppurative, or it may be accompanied by pus-formation, with perforation of the drum and a suppurative discharge from the ear. *Otosclerosis,* an important disease of this part of the ear, is a chronic affection of the bony capsule which entails a "spongifying" of the bone about the oval window, and may lead to a fixation of the stirrup-plate. Both deafness and subjective noises are associated with this affection, which is often hereditary, and frequently involves both ears. With this disease, low-pitched sounds are hardest to hear, while the higher-pitched sounds often remain normal. There is a pronounced shortening in the duration of hearing by air-conduction at the lower end of the scale, as compared with sounds conducted through the bones of the skull. Subjective noises and those like a rushing stream are frequent, and the patient is often able to hear better in the presence of noise than when all is quiet. The presence of many sounds of considerable intensity has the effect of overcoming in some measure the inertia of the stirrup, thus forcing it into activity. Once this is made to function, other sounds of weaker intensity become audible.

The occlusion of the Eustachian tubes is a frequent cause of deafness. It may be occasioned by catarrhal conditions of the nose and pharynx, and by adenoid growths. Children are especially susceptible to this disease, which usually affects both ears. With adults, crusts may form at the opening of the tubes to cause a stoppage, and thus disturb the balance of air-pressure on either side of the drum-membrane. The patient becomes hard of hearing, and also is likely to experience subjective noises and the resounding of his own voice (*autophonia*). Under these conditions, sounds, normally inaudible, which arise from the flow of the blood and from contraction of the muscles, are transmitted and perceived through the sound-conducting apparatus, because of the increased tension of the drum-membrane occasioned by a one-sided air-pressure.

With the closure of the tubes, serum may collect in the middle ear, followed by the formation of pus. More often an acute affection of the nose or naso-pharynx extends through the tubes to the middle ear. As a result of suppuration here, the drum-membrane may be perforated, or the suppuration may extend through the round window into the lower coils of the cochlea. Any interference with the sound-conducting mechanism of the middle ear results in deafness, especially for the lower range of tones. A chronic impairment of hearing may follow repeated attacks of inflammation in the middle ear, when adhesions result between the ossicles and the walls of the drum-cavity. Defects of hearing and low rumbling noises are also associated with motor neuroses which involve the stapedius muscle and the *tensor tympani*.

The diseases of the inner ear are attributable to direct infection or to *toxæmia*, mechanical injury superinduced by very intense sounds, degeneration of nervous elements, histological changes of the cochlear membranes, etc. So long as low tones are heard, the defect can be safely located in the inner ear, except in the cases of otosclerosis already mentioned, which involve the capsule of the stirrup, and in cases of acute exudations in the middle ear, wherein low sounds mediated by bone-conduction are always prolonged and increased in intensity.

A disease of the inner ear may involve any part of the scale, but when the defect is definitely circumscribed, a labyrinthine affection is indicated. Nervous deafness, however, may be attributable either to the cochlea or to the nervous system. Subjective noises are an inadequate criterion for the location of the disease, since they occur in both middle- and inner-ear affections, yet a *tinnitus* of inner-ear origin is usually of a high whistling pitch, while one attributable to disease of the middle ear has a low, rumbling, rustling, character.

Inflammation of the labyrinth is directly caused by meningitis and syphilis, or may follow in the wake of scarlet fever, influenza, typhoid fever, mumps, measles, osteomyelitis, small-pox, and whooping-cough.

Bezold (*ibid.,* 299) maintains that half the cases of acquired deaf-mutism are attributable to cerebrospinal meningitis. 'When occasioned by syphilis, deafness may be either hereditary or acquired; the prognosis is always bad, though better in the acquired than in the hereditary cases.

Disease of the auditory nerve may result from typhoid fever, tuberculosis, and from various constitutional ailments. Certain toxic substances, such as quinine, salicylates, etc., also occasion temporary dizziness, ringing in the ears, and deafness. Degeneration of the auditory nerve is superinduced by a diseased condition of the surrounding tissues, such as may be caused by a spongifying of the bone or by cerebral tumors. Mechanical injuries of the ear are of various sorts, such as those of surgical operations, self-inflicted injuries, the effects of gun-shot, and fractures of the skull. The detonation of an explosion may cause permanent or temporary deafness; so may continued exposure to the loud noise of machinery or the rumbling of locomotives.   Engineers, boiler-makers, and firemen are frequent sufferers on this account.   An interesting comparative study of the degeneration of the cortex produced in guinea-pigs after a constant stimulation by jars and sounds has been made by Karl Wittmaack (*140*).   He found that, after prolonged jarring, degeneration set in at the apex of the cochlea, and after prolonged stimulation by sound, at the base.   From this he concludes that deafness produced by auditory stimuli is associated with the higher range of the scale, whereas deafness resulting from jars and bodily shocks applies to the lower.

A special type of occupational deafness is found among workers in caissons, who may be subjected to sudden changes of air-pressure.   Unless the air-pressure is equalized in the ear by keeping the Eustachian tubes open, the drum-membrane may be ruptured, and the action of the middle-ear mechanism disturbed.

DEAF-MUTISM

While deaf-mutism is often inherited, Bezold estimates that in more than half of the cases he has studied it was based upon an acquired deafness.   Deafness may arise from anomalies in development during the embryonic period, or may result from general diseases, or local disease of the ear, during early life.   About one-third of all Bezold's cases indicate a remnant of hearing sufficient to allow the patient to use his ears if he is properly educated to do so. Boys appear to suffer more frequently than girls from deafness produced by disease, while girls are more frequent among the cases of the congenitally deaf. Although geneticists seem to agree that deaf-mutism is a heritable trait in certain families, apparently it is not attributable to a simple unit-character of the recessive type, since cases are known in which the offspring of parents both of whom were deafmutes have possessed normal hearing. Yet with consanguinity of the parents the chance that the defect will be inherited seems to be greater than otherwise. The difficulty of determining with certainty whether a given case of deafmutism is inherited or acquired complicates the problem of tracing the appearance of the defect through successive generations.

Since deaf-mutes are not lacking in a capacity to speak, an appropriate training will develop this faculty. More significant for us is the means of training a partially deaf-mute to make use of the remnant of hearing which he possesses. The method of such training is only indicated after a careful examination of his ability to hear. If this can be determined, our knowledge of the vocal sounds and their formants can be used to suggest appropriate exercises in the perception of speech-sounds.

Often a child whose hearing is defective fails to take an interest in the sounds of the human voice, although he can be led to do so when those that are audible to him are practised until their linguistic significance is understood. When certain vocal patterns are correctly apprehended, others, even when incomplete or lacking, can be allowed for in the total flow of speech. The individual thus learns to comprehend the spoken word, and can manage to make out what is being said in his presence. A good deal of success has attended special training with the vowel-sounds, which are so important in the scheme of vocal patterns. These are presented to the patient by artificial or by natural means in brief practice-periods which are repeated until he is able to apprehend most or all of them by reference to their characteristic formants. Upon these fundamental patterns of perception the syllables are then built up until he is in command of whatever range of vocal sounds his scanty hearing will allow.

PSYCHOGENIC DEAFNESS

A quasi-deafness may be entirely psychogenic. Although the sense of hearing is intact, the individual makes no attempt to comprehend what he hears. He may even be unaware of sound, so definitely is his attention distracted therefrom. Cases of this kind are supposed to be associated with infantile fear, sound being one of the natural stimuli for instinctive fear-reactions (*131*, 195). If a definite "complex" has been built up in early childhood leading to the avoidance of response to sounds in general, or more usually to certain sounds in particular, such as those of the human voice or of musical instruments, the problem is primarily that of relieving the individual of his obsession. Since the complex is a significant pattern associated with an avoidance of the usual responsive reactions to these sounds, a cure may sometimes be effected by psychoanalysis. When the physician brings to the patient's mind the meaning of his fears, and shows them to be groundless, or at least capable of rationalization, the obsession vanishes, and the patient begins to take interest in speech or in music, and may thus be educated to perceive quite normally sounds he has hitherto involuntarily refused to apprehend.

Amusia, or deafness to music, is frequently met with. The avoidance of effort to comprehend musical sounds passes with less comment than does deafness to the spoken word. Consequently, the occasion for this lack of interest may be less morbid in nature than that in which one refuses to comprehend speech. Bernfeld (*11*) has described two cases in which an amusia appeared to be grounded in a psychogenic complex, but it would be hazardous to assume that a *phobia* is at the root of all cases where individuals of an otherwise normal hearing fail to show an interest in music. Music, like oral speech, must be understood in terms of its own patterns and their logical arrangements. Were it as inevitable for the normal human being to understand music as it is for him to understand oral speech, we should probably find fewer unmusical persons than we now do. Since music is much less closely associated than speech with the practical exigencies of life, many persons are *amusic* because they never have been properly educated to hear music.

A perceptual disability is not the only cause of amusia, any more than it is of word-deafness. Pathological derangements affecting the attributes of sound may be responsible for both conditions. In music this is peculiarly obvious when we consider the integrative patterns of pitch and volume upon which the phenomena of tonality, fusion, and musical interval are based. Regional defects are also of significance, as was found to be the case with von Liebermann (*96*).

MECHANICAL AIDS TO HEARING

Whenever deafness is incurable by therapeutic or operative means, and the limits of re-education applied to the remnants of hearing have been reached, certain mechanical devices may be applied to improve the conduction of sound, and to increase its resonance in the ear. Ear-trumpets and acousticons often improve hearing, and certain electrical devices may help cases where defective hearing is occasioned by impairment of the middle ear. They do not, as a rule, effect any improvement where the disease involves the inner ear or a nerve-lesion. Even when applicable, the difficulty of securing a satisfactory adjustment of these instruments to the affected ear is sometimes insurmountable. Since the defects of hearing are likely to be greater in some regions of the scale than in others, and since the sensitivity to different types of sound may also vary, the adjustment of the instrument so that it will reinforce certain

sounds and not others is no easy matter. Great care must
be taken to secure a mean intensity of the sounds magnified,
so that special zones of favored intensities may be relied
upon to help rather than to hinder the patient; one must
also seek to avoid the often distressing reverberation which
attaches to certain sounds when modified by the apparatus.
An exact and painstaking diagnosis supported by a more
scientific knowledge of auditory defects may in future lead
to the construction of differential devices of a mechanical
nature, which will improve the hearing of deaf persons much
more effectively than has hitherto been possible.[1]

SYNÆSTHESIA

Among the anomalies of hearing there is one which, while
probably not of pathological origin, is sufficiently unique to
be mentioned in the same general category with amusia.
This is the persistent association of certain colors with tones,
vowels, words, and even with certain persons, hours of the
day, days of the week, months of the year, etc., which is
termed *synæsthesia*. We shall confine our reference to the
association of color and tone, chiefly with the purpose of
warning the reader that such fragmentary investigations as
have been made of this curious anomaly do not warrant the
opinion, sometimes advanced, that a universal one-to-one
correspondence obtains between the color and the tonal
series.

[1] Cf. *150.*

While a person possessing a synæsthetic capacity may be
fairly consistent in imagining or thinking certain colors with
certain tones, thus showing the habit to be deeply ingrained,
if not precisely instinctive, great variation is found when we
compare the experiences reported by one synæsthetic person
with those of another. The only trends that are at all frequent seem to be those of associating either tints or colors
of short wave-length, like blues and violets, with tones of
high frequency, and shades or colors of long wave-length,
like reds and yellows, with low tones (75, 237). But even
these parallels between the brightness-series or the color-
series of vision and the series of tones are not universal;
for C. S. Myers reports a case (77, 115) in which a tone of
256 v.d. was blue; 400 v.d., violet; 500 v.d., red; 600 v.d.
and 700 v.d. indefinite, the latter perhaps a light green;
while 800 v.d. and 900 v.d. were blue again, and 1,200 v.d.
and 2,048 v.d. were very translucent and yellowish.

The color may also vary with the timbre of the instrument
employed in giving the tone. In the rather striking case of
the well-known Russian composer, Alexander Scriabin, the
colors which he "heard," and which he introduced as a natural and desirable accompaniment to some of his musical
compositions, are associated neither with tones nor timbres
per se, but with *tonalities*. To quote a paragraph from
Myers' study of the case: "The strongest colors for Scriabin
appear to be those relating to the keys c major, d major, b
major, and f♯ major, placed respectively in red, orange-
yellow, blue, and violet. Starting, however, from c at the

red end of the spectrum, Scriabin finds that as he passes from hue to hue, the successive colors correspond to tonalities rising by a series of fifths. Thus the key of c is red,

Red	Orange	Yellow	Green	Blue	Violet
c	g	d	a	e	b	f♯

of g red to orange-red, of d orange to yellow, of a yellow to green, of e green to blue, of b blue to violet, and of f♯ violet. The colors of the remaining keys are believed by Scriabin to be extra-spectral—either ultra-violet or infrared. Thus the key of f is 'on the verge of red,' giving often the effect of a metallic lustre." (*Ibid.*, 114.)

There appears to be no reason for the assumption that such an arrangement is other than arbitrary, yet a "color-music" in which the principles underlying gradation and contrast are appropriately developed is by no means impossible; and its association with the music of tones might not be unworthy of serious artistic effort.

The bond of "sympathy," as Myers terms it, whereby synæsthesias are established, perhaps in early childhood, is a matter of some psychological interest, but the cases thus far studied have been too few in number, and too varied in character, to afford a satisfactory clue to the origin of the phenomenon.

MELOTHERAPY

In concluding this chapter on the pathology of hearing, brief mention should also be made of the curative effects sometimes claimed for music, especially in the treatment of the insane. While it is undeniable that music often soothes an excited person, and may therefore be beneficial as a means of treatment in certain psychopathic and neuropathic cases, even in such cases its beneficial effects seem to be conditional upon the musical capacity and aptitude of the patient (*23*, 177 f.). As in the cases of synæsthesia, so here, there is no one-to-one correspondence between certain sounds and certain ailments, either of body or mind, upon which, with our present knowledge, we can construct a rational method of treatment. Melotherapy, like the art of colored music, is but a blind groping after something, the very existence of which is highly uncertain. What we may infer from these phenomena is the intimacy and the permanence of certain sensory integrations, which may or may not be fortuitous in origin; but the application of these integrations in everyday life is so remote, and so seldom called for, that only in exceptional cases are they definitely expressed and recorded. Since we are at a loss to rationalize a persistent association of color and tone, or a reflective mode of bodily response with a certain type of melody or harmony, these phenomena are apt to be overlooked by us as bits of mental furniture for which we seldom find any use. Fortuitous and individual though they seem to be, the readiness with which these integrations are formulated may be congenital. Even the so-called "conditioned reflexes," in which a response is experimentally transferred from its original stimulus to another of a quite different order, raise the question of integration. When Pawlow showed that dogs could be trained to secrete saliva when a bell was sounded, if in the course of their training this sounding of a bell had invariably accompanied the sight and smell of food, we were, perhaps, too ready to generalize from this experiment, and to conclude, not only that any response may be called forth by any discriminable stimulus, but that the substitution of stimuli is effected without any "insight" on the part of the animal. The association of the sound of a bell with a watering of the mouth is quite as irrational as the association of sheer color and sheer tone, or the apprehension of a certain type of melody with a certain bodily resonance conducive to healthful functioning. Yet it does not follow that any of these connections are external or arbitrary in their mental and bodily operation. The integration of the attributes of consciousness to form the phenomenal structures that constitute our varied experiences of sound may be judged equally irrational, but these integrations certainly do not result from external combinations of partial ingredients arbitrarily brought together. On the contrary, the very essence of an integration is a creative synthesis, and such creative syntheses may, and in all probability do, underlie the conditioned reflexes, just as they underlie the more usual responses of a hungry animal in the presence of food. All these cases are subject to the laws of perception, and every anomaly of animal behavior, whether it be mental or physical, functional or pathological, must be considered in the light of structural principles of the kind we have been able to set forth in the course of this book.

In the practice of medicine empiricism has always held sway. Since we possess so little knowledge, our diagnoses are necessarily crude, and our therapy groping; yet every marked improvement in the technique of the practitioner rests upon some new insight into the nature of his patient's ailment, and in this an understanding of the mental phenomena that reveal so many of the symptoms of disease is quite as important as the anatomical and physiological data with which the medical man so often is content to rest. The pathology of hearing is not exclusively founded upon the anatomy and physiology of the ear and auditory nerves; it likewise shares in the results of psychological investigation, and in a study of the principles of mental organization which these results indicate.

CHAPTER XIII

AURAL EDUCATION

THE PRE-EMINENCE OF SPEECH AND MUSIC IN EDUCATION

ALTHOUGH speech and music are by no means the only fields of study into which the perception of sound may enter, they are the ones in which educational methodology attains its greatest significance. When we listen to the sounds of daily life, or observe for scientific purposes the cries of birds and animals, the varied voices of nature, or the noises of human instruments and machines, although acuity of hearing may be desirable, and even indispensable, it is primarily the registration of the sound with reference to the object from which it emanates that most concerns us. In the case of oral language and music, however, the nature of the sound itself is also of immediate interest, and it is through education alone that man is able to define, and thus to dwell upon, these phenomena in their orderly arrangements quite apart from their varied objective references. The child, in learning to speak, to read, and to comprehend the speech of others, is interested not only in the objects that produce the sounds he hears, or the things to which the words refer; he is interested also in the sound itself as a phenomenon in a setting of other sounds and other experiences. Thus we are apt to conclude that many of the sounds uttered by a child when reading aloud, and also many of the sounds to which he gives attention, particularly those uttered by his parents and attendants, are quite meaningless to him. Meaningless they may be, but only in the sense that the sounds of an unknown tongue are meaningless; for whenever attention is aroused, the object of attention must have a phenomenal existence which under appropriate conditions will compel the definition of its contour as a perception. Whenever a sound is perceived, it is at once defined by the primary integration of certain attributes, for only in this way is the hearer able to set it off from other experiences of the moment. Since the sounds both of speech and music are elaborately constructed, they do not so readily pass into secondary integrations with other objects, which can be seen and touched, as do the common noises of our environment. Consequently, speech and music are more favored as material for education than are the sounds one associates with storm and flood, with birds and animals, or with machines and other human inventions. Although, when first heard, these, too, are but sounds without a definite reference to the physical events that occasion them, so soon as the necessary connections have been established, the sound becomes absorbed in its generating cause. Thereafter we listen not so much to the call of the bird as to the bird itself, and not so much to the whirring of the machine as to the machine itself. This attitude is also acquired for the sounds of speech and music, but not so readily, nor so completely; because the patterns of speech and of music possess a more elaborate inherent orderliness than do the sounds of nature or of the man-made environment. Music and speech, accordingly, remain a subject of special interest in their own right, as phenomena in which we can take delight, and in the production and apprehension of which we find many absorbing problems.

In learning to speak and to understand speech, the child must master a number of aural problems and their appropriate solutions, and while the demands of a musical comprehension are less insistent, the materials of music are nevertheless bestowed upon each who has a capacity to hear tones. Music, then, is not less "natural" in its origin than speech; for the fundamental patterns of both are laid down in the perceptual and ideational capacities of the human being.

AURAL EDUCATION IN SPEECH

The problem of aural education in speech is but inadequately solved with our present methodology of instruction, because we lay too much stress upon the attachment of linguistic sound to things and events, and make too little of the structural relations inherent in the simple aural patterns themselves. A good pronunciation, linguistic rhythms, and all the niceties of refined utterance which constitute the basis of "style," or beauty of form, are apt to be neglected at a time when these aspects of language make their natural appeal to the child. If we were ever mindful of the fact that a sound makes its structural appeal by virtue of its definiteness, and that this definiteness has an æsthetic quality of completeness and balance, we could perhaps prevent the development of the strained high-pitched voice, the careless integration of linguistic sequences, the failure to build upon and to elaborate the natural rhythms of speech, and the general lack of discernment in the selection of words and their enunciation which so often characterize vocal utterance. Every child is born an artist; and without rhythm, assonance, and word-tunes, he never could remember and reproduce the cadences of speech. Not only are these cadences dependent upon principles of perception which aid the memory; they are likewise essential to an understanding of the words spoken by others. Yet it is very easy to slur the details of these patterns, to introduce faulty phonations and disturbed rhythms; and while the result necessarily is less clear and less intelligible than it otherwise might have been, so intent are we to comprehend whatever we hear, that we strive to catch the drift of meaning, even when the utterance is very deficient. Thus as teachers we may even encourage the habituation of errors in our pupils by allowing these errors to pass without correction.

A teacher will often jump at the intended meaning of a verbal utterance or composition, and be satisfied when he is able to catch it. But in so doing he stresses the object-meaning of the words, and fails to cultivate the æsthetic elements of the linguistic style of his pupil. This pedagogical error bears fruit in the vain struggle which ensues during the later periods of the child's education to recover by rational means what has been lost in childhood for lack of sympathetic and intelligent guidance.

Children are by nature poets; they delight in speech for its own sake, in rhyme and rhythm, in assonance, alliteration, and cadence. Yet teachers permit these formative elements to slip into the background, thenceforth to serve merely as the bare necessities of a framework upon which the ideas of objective reference may be built. No wonder we grow up with so slight a command over the instrument of our speech that, without the toilsome study of an arid grammar supplemented by the comparative examples of a foreign tongue, we never after are able to control our language well enough to carry on a decent conversation, or to compose a significant paragraph. Though we study "language" throughout the twelve years of the public-school system of instruction, we have not attained proficiency enough to pass muster in the college, but are forced to study English at least another year; while some special merit seems to attach to those who find it desirable to continue the study throughout their entire course of higher training. Not that linguistic study may not be a profitable venture; its problems are numerous, and its rewards significant. But we might know, most of us, how to speak and write acceptably much sooner and much more readily than we now do, if our linguistic education began with the native impulse to speak and write. Yet this impulse is deliberately neglected by teachers in order that the child may first acquire a purely external vocabulary of signs and tokens.

By a mistaken method of instruction, children are permitted, when they are not positively encouraged, to neglect the "feel" for linguistic sound, at a time when the perception of these sound-structures and their relations is quite as significant to them as are the references of words to objects and events. Thus a large part of their later education must be directed towards recovery, by the rational means of grammar, syntax, and phonetics, of something that was originally theirs by instinct.

The more desirable course of instruction to which we refer has a far wider application than that of reference to speech alone, for it proceeds from a knowledge of the rhythms that underlie all modes of expression, and likewise of the phenomenal structure of every perceived object. One phase of the whole subject is being developed by the method known as *Eurhythmics,* of which the well-known musical educator, Émile Jaques-Dalcroze (*43*), is the leading exponent. But if we trace back the origin of *Eurhythmics* to the education of the ancient Greeks (*79*), we shall find that the method had for them a still broader connotation, being inseparable from the general methodology of instruction in speech and the graphic arts. The Greeks recognized that all true education is one in which both instinct and tradition co-operate; for the forms of expression are rhythmical in their sequence, and the forms of apprehension are patterned and designed. To these ultimate forms, then, attention must be drawn by appropriate methods of education.

For our present purpose, it is the auditory data of experience that concern us; but we dare not forget that visual, tactual, and kinæsthetic elements are inseparably bound up with the apprehension and expression of sound. As we grow older, and have learned to reflect, the method of our education shifts from the immediate data of our sensibilities as employed in the kindergarten to the abstract terms of reason. And here is no doubt a gain; for, as Dalcroze remarks: "Reasoning is so much simpler than imagining" (*ibid.*, 139). Yet "the admirable offices of logic in concatenating ideas are, however, a poor substitute for the rapid processes of spontaneous emotions, creative of vivid images, thrown out haphazard in their violent course, forming in their fortuitous groupings ever new effects of combination. No adequate attempt is made in our schools to develop children's imaginative qualities; on the contrary, these are suppressed by a continual insistence on analysis, co-ordination, classification, and labeling." Whenever the emphasis is turned too quickly from the contemplation of the structure of words to the meaning of the objects or events to which these words refer, this labeling process is subversive, for, without adequate envisagement and an ability to enjoy the sounds themselves in their sequential patterns, we are lacking in control of speech and in the subtler means of expression. "I contend," says Dalcroze (*ibid.*, 125-6), "that schools ignore the training of sensibility, with deplorable results upon the development of character. It is—to say the least—strange that, with the existing prevalence of neurasthenia, no attempt is made to direct the boundless desires arising from ill-controlled feelings; that, in newly-developed countries, where, for the most part, will-power is concentrated . . . upon the attainment of commercial success, educationalists show no anxiety to awaken the moral sense of the coming generation, while in countries where traditions too long established have a cramping influence on the development of individuality, no resort is made to expedients for arousing temperament. And yet the means are at hand whereby the coming generation might be trained to a greater flexibility of spirit, a firmer will-power, an intellect less dry and exclusive, more refined instincts, a richer life, and a more complete and profound comprehension of the beautiful."

Instruction in speech should be directed first of all to the conservation of that interest in sounds themselves and their utterance which every child possesses as a birthright. While we must not neglect the reference to objects and events, yet the æsthetic qualities of sounds, and their articulation, should ever be foremost in the mind of the teacher of linguistics. Among scientific aids to expression, a knowledge of phonetic principles is very helpful, since children often find difficulty in the pronunciation and apprehension of certain of the important sounds of speech. Knowledge of the mouth- and throat-positions requisite to these difficult articulations will suggest means for securing them. Exercises in breathing are also helpful, and the various methods of speech-correction should be understood, so that undesirable habits such as stuttering and stammering may be

checked at the outset, and the baneful effects of vocal dis-
abilities, from which so many persons suffer, may be avoided.
While intelligent effort will find a means to overcome some
of these hindrances even when they persist, no one can doubt
that the handicap of an imperfect command over the instrument of speech is always serious. The enjoyment derived
from the satisfaction of control over speech goes a long way
towards making life tolerable; for an assurance that we can
express ourselves adequately encourages us to think broadly
and deeply.

AURAL EDUCATION IN MUSIC

In turning now to the pedagogy of music, we enter a field
which, in comparison with speech, has been allowed to become more abstract and remote from everyday life; for an
intelligent person must strive to speak and to comprehend
the words of others, but no constraint is put upon him
either to sing or to enjoy music. The Greeks understood the
function of music in life better than we do, and although
their music was simpler than ours, its intimate bearing upon
speech and bodily movement gave it an integral part in
life which it now lacks. Music has become something
that concerns only those with peculiar talents, who learn
it as a trick is learned in order to furnish them with a
special accomplishment. But "actually," as Dalcroze has
written (*ibid.*, 82), "music, like gymnastics, is primarily not
a branch of learning, but a branch of education. The school,
before everything else, should aim at molding the physical

and psychic personality of the child, at preparing him for life. If we postponed the study of Roman history till we had reached twenty, our general development would not be affected. But to commence our gymnastic exercises and music practice at an adult age would be to lose most of the benefits they should provide. Gymnastics mean health; music means harmony and joy. Each of them provides a refuge and a reaction from overwork. To make students sing daily, if only for a quarter of an hour, would be analogous to setting them every day, between each two lessons even, a few physical exercises. The singing of ballads and songs would thus become a natural practice with school-children, while singing and music-lessons would be made to form a part of the school curriculum. These should be devoted to the study of musical science, and should be in proportion to the other branches of learning. They should inculcate a knowledge, not of singing, but of music and how to listen to it."

To quote next an American musician and writer who has given much thought and effort to the promotion of musical education among us, Thomas Whitney Surette (*128*, 39): "The prime object . . . of musical education for children is so to develop their musical sensibilities as to make them love and understand the best music. Does this bring up the question, 'What is the best music?' By the 'best' music I mean exactly what I should mean if I were to substitute the word 'literature' for 'music'—I mean the compositions of the great masters. And if you say that the great masters did not write music suitable for little children, I reply that such music has nevertheless been produced by all races *in their childhood,* that it exists in profusion, that it is commonly known as 'folk-song,' that it is the basis upon which much of the greatest music in the world rests, and, finally, that it is the natural, and indeed, the inevitable means of approach to such great music."

The "folk-song" provides an appropriate content of musical instruction for the same reason that "tradition" was the approved content of Greek education; for whatever has been achieved by the slow process of a natural evolution must possess in a high degree those elements of formal structure which are both native and felicitous. The folk-song grew up as a communal expression, just as have certain dances, ceremonials, poems, and legends. In adhering to these as a basic content of education, we are guaranteed in advance that their structure is appropriate both to the apprehension and to the means of expression which the child possesses.

Unfortunately, musical education in our homes and in our schools is not pursued in this spirit, and may, indeed, be altogether neglected. While a higher education must, perforce, endeavor to supply by rational means something of the command over spoken forms which has been allowed to slip away through neglect in childhood, it is impossible to impose stringent devices in enforcing a study of music upon the adolescent. If he has failed to catch the spirit of music, or to elaborate its forms for his own amusement before he turns twelve, it is unlikely that he will greatly profit from an enforced régime of musical instruction thereafter.   Consequently, one of the practical problems of musical education is that of selecting individuals whose talent, native or acquired, is sufficient to warrant them in the expenditure of time, energy, and money upon their music-lessons.

TESTS OF MUSICAL TALENT

At present an immense amount of time and money is wasted on the musical education of children.   It has been estimated that in this non-musical country of ours, musical education, so-called, annually absorbs millions of dollars in excess of the total expenditure upon all other forms of education combined (*104*).   That the apparent results are so meagre, in the cultivation of both musical appreciation and performance, is perhaps the best ground for the imputation that we Americans are, indeed, an unmusical people. But whether the censure be true or false, it seems reasonably clear that a great deal of misdirected energy and preventable waste is involved in the teaching of music.   The effort is misdirected if the individual is incapable of profiting by his instruction, and the waste is preventable if it is possible to test a ten-year-old child with sufficient accuracy to determine in advance whether or not he is likely to improve his opportunities in the field of music.

The solution of the problem has been sought through vari-

ous forms of examination, the results of which may be expected to yield trustworthy information concerning the nature of one's musical talents and their educability. Many
psychologists and educational experts have accordingly engaged in devising and standardizing simple tests of musical
capacity and ability—tests which can be easily given, and
whose results can be relied upon. These tests, for the most
part, are empirical rather than scientific. On the one hand,
the phenomenal structures which underlie musical thought
and its expression have not been sufficiently understood to
make possible an exact estimate of individual talent, and,
on the other hand, the conditions under which the tests are
made lack scientific precision. Instead of an accurate estimate based upon scientific observation of the individual
examined, we must rely mainly upon judgments of common-sense, which have scientific validity only with respect
to the correlation of a child's performance in the tests with
certain opinions of his talent otherwise secured. One can
perhaps never expect a high degree of scientific accuracy in
the observations of a ten-year-old child, but when one un-
derstands the compulsory conditions under which the phenomenal structures of musical thought and appreciation occur, one can at least control these conditions to a considerable extent in performing the test, so that the results of
examination may be expected to show a closer correlation
with the individual's real musical capacity than has been
hitherto obtained.

Among the many investigators who have turned their at-

tention to this problem, Carl E. Seashore and his associates
at the University of Iowa have for many years been engaged
in devising and standardizing a variety of musical tests.
With the aid of an ingenious instrument which he calls a
"tonoscope," Seashore is able to obtain a visible record of
the sound produced by the human voice, or by any other musical instrument (*107*). In this way, a singer may see before
him a record of his own vocal ability, and learn how much
he tends to "sharp" or "flat" in a certain region of the scale.
The record of his performance is a useful means of ration-
ally directing his efforts towards a correction of his errors.

Seashore has also devised, and is standardizing, a variety
of tests for determining the presence or absence of different
elements of musical talent. We shall consider these tests in
some detail, because the problem is important; but we shall
be obliged to take exception to certain of the conclusions
Seashore has reached, since his tests do not appear to take
into account all the attributes of sound upon which musical
ability rests.

The Seashore tests are classified in four groups: (1) sensory tests of ability to hear music; (2) motor-tests of ability to sing; (3) associational tests of ability to remember,
imagine, and think in terms of music; (4) affective tests of
ability to feel music. Each group consists of a number of
different tests. The first two groups are concerned with
pitch, intensity, and time. "The elements of musical sound
are really three," writes Seashore, "namely, pitch, time, and
intensity. The fourth attribute of sound, extensity, which

represents the spatial character, is negligible for the present purposes" (*105*, 130). To the last statement we must, of course, take exception, but our criticism will follow later. In the third, or associational group of tests, imagery, memory, and ideation are the points covered; while the fourth group of affective tests deals with the character of the musical appeal, the emotional reaction, and the power of æsthetic interpretation.

Among his tests, Seashore regards the one for discrimination of pitch as in a measure fundamental. Not that all who can discriminate pitch with accuracy are necessarily musical, but that "this test is basic in the sense that many other aspects of musical capacity rest upon the capacity here measured. Thus, tonal memory, tonal imagery, the perception of timbre, singing and playing in true pitch, and, to a certain extent, the perception of harmony and the objective response, are limited by any limitation that may be set in pitch-discrimination. The evidences for these facts are the result of both theoretical analysis and cumulative experimental record. If the pitch-discrimination is poor, we can predict, at least, a corresponding inferiority in the derived factors. On the other hand, excellence of pitch-discrimination does not necessarily insure excellence in these factors, since it is only one element in them" (*106*, 65-6). His conclusion is supported by tests of a large number of persons, of different ages and different sex, the results of which indicate a certain normal distribution in the ability to discriminate pitch. A frequency-curve for 1,265 university students, reproduced in Figure 18₁(*ibid.*, 47), shows approximately 1 per cent able to discriminate ¼ of a vibration about a standard tone¹of 435 v.d.; 12 per cent, 1 vibration; 31 per cent, 2 vibrations; 23 per cent, 3 vibrations; 14 per cent, 5 vibrations; 9 per cent, 8 vibrations; 4 per cent, 12 vibrations; 3 per cent, 18 vibrations; 2 per cent, 25 vibrations;

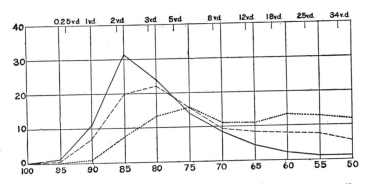

Fɪɢ. 18. Distribution of Capacities in the Sense of Pitch according to Seashore's Measurements. Solid lines, adults; dashes, eighth-grade children; dots, fifth-grade children. The numbers at the bottom denote the percentage of right judgments; and the numbers at the left the percentage of cases at each level. The distribution of *limens* is indicated on the horizontal line above.

and 1 per cent, 34 vibrations. Similar curves are given in the figure for fifth-grade and eighth-grade children. Although the distributions are similar, it will be observed that the threshold for the largest percentage of eighth-grade pupils was about 3 v.d., and of fifth-grade pupils between 5 and 8 v.d. A comparison of these normal distribution-curves of adults and children also indicates the discrimination of children to be less uniform than that of adults.

"A successful experimenter," writes Seashore, "can reach

the proximate physiological threshold in this measurement within a half-hour's trial in about three-fourths of all cases. In the remaining one-fourth, he must either suspend judgment, or proceed with special methods to eradicate the difficulty, which can be done with all persons reasonably well-developed mentally, although the process often requires great patience and ingenuity. As a rule, those who have keen ears for pitch-discrimination reveal the physiological limit at once, regardless of age, training, or general brightness" (*ibid.*, 51-2).

Seashore concludes, therefore, that he is dealing with an elemental capacity whose physiological limits are readily approached. The difficulties that have to be removed before one can reach an accurate judgment are found to be of a cognitive nature, such as "ignorance, misunderstanding, inattention, lack of application, confusion, objective or subjective disturbances, misleading thought, inhibitions in recording, etc." (*105*, 136). In most cases these are so readily set aside that the sheer physiological capacity of the organ of hearing is quickly approximated, and no subsequent training is found to have any appreciable effect in lowering the threshold that is then reached. The evaluation of these results as a prognosis of musical talent is given as follows: Those who discriminate better than 3 vibrations with 435 v.d. as a standard may become musicians. Those whose discrimination falls between 3 and 8 vibrations should have a plain musical education (singing at school may be obligatory). Those whose discrimination falls between 9 and 17 vibrations "should have a plain education in music, only if special inclination for some kind of music is shown" (singing in school should be optional), while those whose discrimination is 18 vibrations and above "should have nothing to do with music" (*106*, 66-7).

As regards the general value attributed to this test, several things may be said. In the first place, the diagnostic worth of a test in pitch-discrimination which makes no allowance for differences of volume or the attendant effects of interval must be seriously questioned. It is evident, of course, that one whose sense of pitch is defective will lack a full appreciation of musical intervals; yet the judgment of volume is an even more important factor than that of pitch, as was indicated by the anomalous case of von Liebermann, whose sense of consonance was impaired in certain regions of the scale, while his ability to discriminate larger and smaller intervals remained intact. Indeed, a contrary condition might well be possible in which one would be accurate in judging intervals, yet very inaccurate in judging pitch, though experimental evidence upon this point is still wanting.

In connection with the evaluation of these tests, the correlations of numerous results obtained in Seashore's laboratory are instructive, because they are all so surprisingly low (*108*). The correlations of results obtained from a variety of tests with the results of pitch-discrimination were rarely as high as .3. Those that exceeded this figure were the correlations of pitch-discrimination with musical experience, .31

and .38, pitch-discrimination with consonance, .33, and pitch-discrimination with tonal memory, .52. When we remember that a correlation is commonly regarded as significant only when it is above .5 (perfect correlation being indicated by 1.0), we see that the bearing upon one another of the various functions that Seashore has tested can not be very great, or else the particular tests were not altogether trustworthy. While nearly all Seashore's correlations were positive, most of them were low. One remarks, for instance, the correlation of pitch-discrimination with singing the key-note, which was .14 in one group of tests, and .21 in another; also the correlation of pitch with the singing of intervals, which was .15 in one group of tests, and .12 in the other. These results clearly indicate the presence of important elements other than pitch in determining the ability to sound a tone accurately and to establish an interval. Motor-ability is doubtless one factor which here plays a part, but the sense of interval must also be reckoned with.

In dismissing the attribute of extensity as "negligible for the present purpose," Seashore overlooks both the experimental and the theoretical importance of volume as an aspect of sound. While he is "at present inclined to believe that there is such a factor," and states that, "entirely apart from association with other senses or with intensity, timbre, or duration, there is, in the hearing of all tones, an inalienable experience of bigness," yet this "bigness" is erroneously assumed to vary directly with pitch, and is also confused with brightness; for in the same paragraph Seashore remarks that "the high tone is heard as small, piercing, and thin, whereas the low tone is heard as large, massive, rolling, or at least blunt" (*106*, 164).

We have found it possible to formulate a conception of music as an integration of five attributes, pitch, intensity, duration, brightness, and volume, only the first three of which have significance to Seashore. With no intention of depreciating Seashore's valuable work in other respects, we can not fail to note the discrepancies which his own measurements reveal by their failure to correlate significantly with one another. It will therefore not be out of place to remark that a test for the discrimination of intervals should prove of greater value in the prediction of musical talent than a test of pitch-discrimination, for the simple reason that acuity in hearing pitch is musically far less important than accuracy in establishing and judging simultaneous and successive intervals.

Experimental results have shown that the threshold of volume-discrimination in the middle range of the scale, employed by Seashore in his tests of pitch, is a fraction varying between two one-hundredths and three one-hundredths of the vibrational frequency. This would mean, if our interpretation is correct, that the smallest *interval* readily discerned above a tone of 435 v.d. is a tone of about 448 v.d.—an increment of some thirteen vibrations. According to Seashore, one who judges no better than this is musically incapable. The verdict is probably correct if a person can get "off the key" to such an extent without knowing it. But it does not necessarily follow that he would do so; for even when the estimate of pitch is very inexact, the judgment of an interval can be corrected by reference to its volumic proportion. Thus there is a possibility that some of those persons whose pitch-discrimination seemed so defective to Seashore were not judging pitch at all, but were intent upon establishing intervals; and if this be true, it would account for the peculiarly "skewed" character of Seashore's curves of distribution. Only a small percentage of his adult observers discriminated pitch by less than one vibration. The largest number required approximately two; and there follows a gradual decrease in the percentage of observers who could discriminate three, five, and eight vibrations. The gradual slope of the right half of the curve (see Fig. 18) is noticeable not only for adults, but even more for children; with fifth-grade pupils the curve is flat between 8 and 12 v.d., and then actually rises. Nearly half these pupils required a difference of more than 8 v.d. in order to establish a threshold. From this point, all three curves tend to flatten out in a manner that indicates rather forcibly a change of attitude, attention, and interest on the part of the observers. May we not infer, then, that of those who estimated pitch so poorly some were not judging pitch alone, but a complex of attributes among which volume was an important element?

These are the reasons for our general conclusion that the examinations made by the Iowa psychologists might be improved through the introduction of tests for volumic discrimination and interval. It should not be supposed, however, that Seashore relies entirely upon the test of pitch. In the Psychology of Music Studio at the Iowa State University, the individual is subjected to a variety of tests whose results are all recorded graphically upon a chart of musical talent. This chart expresses one's musical ability in many ways, each reckoned upon a percentage basis. The categories of the different capacities measured include: sense of pitch, sense of intensity, sense of time, sense of consonance, acuity of hearing, auditory imagery, motor imagery, motility, time-action, rhythmic action, singing the key, singing the interval, voice-control, the register of the voice, the quality of the voice, etc. The complete results of such an examination give, indeed, a fairly extended "clinical" picture of a person's musical talent. It is evident, of course, that the tests are not of equal value, and therefore it is deemed unnecessary to secure a high rating in each capacity in order to give promise of a talent worthy of cultivation. A musical composer may be an indifferent performer, and a capacity to think effectively in tones is not always combined with a high grade of rhythmical ability; nor is the control of one's voice an accurate indication of musical talent in other fields of thought and expression. Yet, from the large number of tests that have been made, certain norms are obtainable, which are significant in judging individual promise in musical education. But there can be no doubt that the elaborateness of these examinations will be greatly simplified as soon as we are able more directly to test the presence and the functioning of those fundamental configurations of apprehension and response without which a musical talent can not manifest itself.

Other notable attempts to diagnose musical talent have been made by Stumpf (*122*, II, 142) in his studies of tonal fusion as experienced by musical and unmusical persons; unmusical persons tend to accept simultaneously fused intervals without analysis, while the musical person also recognizes the interval involved. Rupp (*98*) has likewise planned an elaborate series of tests, analogous to those employed by Seashore.[1] More recently, Walter Kühn (*53*) has suggested a means of testing musical ability with reference to the feeling of tonality. Kühn's contention, based upon the examination of a considerable number of musical children, is that the fundamental or key-tone of a simple melody in which this tone itself does not occur comes spontaneously to mind when the melody in question is heard. The children were instructed, after hearing the melody, to sing any low tone that occurred to them. In one set of 228 tests with thirty children, 78 per cent of the low tones sung were fundamentals of the melody just heard. This test, however, has not yet been administered to an unselected group of persons, and many of the results obtained with children able to repeat vocally a simple folk-melody, after once hearing it played on the piano, were equivocal, as evinced by the relatively large number of low tones in the intervals of a

[1] Cf. also Pear (*86*).

second or a seventh from the key-tone. We can not be over-sanguine, therefore, that a test of this kind will prove a reliable indicator of musical ability.

In fact, a simple trustworthy test of musical talent has not yet been found; but that it will be found is highly probable; because, after all, the complexity of the musical pattern is not so great that one should be unable to discern its essential constituents. Appropriate tests based upon musical intervals, simultaneously and successively employed with reference both to apprehension and expression, ought to furnish a satisfactory diagnostic measure, which in the hands of a capable examiner would tell us a great deal about the musical capacity of any child. But in planning and experimenting with such a diagnosis we must bear in mind that musical talent is clearly twofold. First, it involves an ability to perceive and to think in terms of tones and intervals, as presented simultaneously, and in rhythmical succession. Secondly, it comprises an ability to express and to communicate musical thought by means of the voice or some other musical instrument, and also by means of compositions recorded on the musical staff. The first ability does not necessarily imply the second; many persons of refined musical taste and judgment are unable to communicate their musical thought either instrumentally, or by musical notation. An eminent composer may lack the technical facility to play his own music, and a virtuoso may be deficient in the creative intelligence requisite to compose music. All questions concerning musical talent and its education are necessarily complicated by these varied aspects of receptive intelligence on the one hand, and the technique of communication on the other.

The phenomena of musical thought and its expression are likewise complicated. In the first place, musical thought rests upon tonality, which is essentially a sense or faculty for dealing with intervals. An interval in music implies all the tonal attributes: the pitch-salient that dominates the tone; its volume, by means of which its interval from another tone is determined; its brightness and intensity, which influence both the effectiveness of pitch and volume; and, finally, its duration, which gives rise to movement, to measure, and to rhythmical order. Some persons are accurate in pitch and interval, but deficient in the sense of measure and rhythm. In singing, they sound their tones correctly, but the rhythmical pattern troubles them. They slur the rest-periods, and are careless of accents. Even accomplished performers often lack the time-sense necessary for weaving together a musical composition of any rhythmical complexity into a moving pattern without jerks and gaps at the periods of transition. To others, music is little more than a rhythmical pattern. They frequently get "off the key," and they appear to have no faculty either for harmonic or for equal intervals. Yet they understand cadence and the rise and fall of pitch-brightness, for they can ascend the scale and come down again as the music dictates, even though they rarely apprehend the right tone, or fix an interval with precision.

All these elements are involved in a new way when we

consider the ability to sing or play upon a musical instrument; for here one must have technical ability in order to express accurately what one has in mind. A musical technique is very difficult to acquire, and in order to keep at the level of automatic response, a performer must practise repeatedly and for hours at a stretch; for, without a perfect command of technique, the performer is unable to give over his mind in thought to the music he is communicating. In testing musical talent, and in measuring musical ability, all these matters require a special, detailed consideration.

THE GENERAL PURPOSES OF A MUSICAL EDUCATION

The first question that concerned us in musical education was the selection of pupils. We have discussed some of the methods of selection, based upon individual tests of significant features in the apprehension and expression of music. The results of these tests seem to show that some individuals are incapable of profiting by a musical education. The assumption probably is warranted, although we have been unable to accept the decision of Seashore in this matter, because his tests do not appear to lay sufficient emphasis upon the significance of the musical interval. According to our theory, musical appreciation and talent are functional aspects of the attributive integrations of auditory sensation. Since the attributes of sensation are universal human characteristics, and hence are essentially uniform (save for abnormalities of a pathological nature), we are led to the conclusion that no one is by nature devoid of musical sensitivity. All we can affirm is that many children seem to lack an adequate stimulus for the development of their musical interests. Undoubtedly there are functional peculiarities inherent in certain individuals, which are of such a nature that an attempt to develop musical ability to any considerable extent would be futile. Yet we can not regard these defects as resident in normal auditory sensitivity. If elementary education were equal to the task of cultivating creative ability, we probably should find that virtually all children are musically receptive, just as they are verbally receptive; although on the technical side there would always be many who could never achieve success as performers, just as there are many who can never achieve skill in speech and writing.

It is evident, however, that the education of musical receptivity should go hand in hand with musical expression of some sort, since only through expressive communication can the child register his musical thought, and thus fully realize its significance. The cardinal point in musical education is that music ought never to be treated as an abstraction. Although singing is doubtless the most natural and simple mode of expression, the dance with dramatic gestures is a very important adjunct. It is highly significant that primitive music should be so closely bound up with the dramatic dance, as an expression of the cult; for this union shows the manner in which creative imagination is controlled and standardized through practical applications.

Every child should be permitted and encouraged to sing and to dance—to express himself by gestures and by words. Furthermore, he should be led to associate these expressions with color and form, and with representations by means of lines, surfaces, and solids. If for the highly-educated the different modes of art tend to separate into special kinds, or *genres,* the fact remains that the creative impulse towards artistic expression is one and indivisible. Out of the unity of æsthetic thought, with its manifold expressions, the special forms of art have only gradually emerged in the history of civilization. For the child, they are still blended. It is therefore poor pedagogy to constrain the child to express himself within the narrow limits of a single distinct branch of art, such as music, drawing, coloring, dramatic gesture, or the dance, where each is taken as a school-subject by itself. Rather ought we to suggest an idea capable of artistic treatment—a star, a plain, a tree—and then encourage a creative response by means of appropriate words, gestures, tones, colors, lines, solids, etc. As the thought awakens in the mind of the child, let that thought find its own mode of expression. A comparison and an evaluation of different responses would then be in order, the intention of the teacher being to cultivate an appreciation of the forms themselves, while at the same time a technique is being acquired adequate to their execution. If our educational methods were more flexible to start with, we should be better able to reveal and to cultivate the innate art of the child. Music, poetry, representative art, and the dramatic dance would then evolve in unison, enriching the mind and improving the conduct of the pupil.

As regards a specifically musical education, our traditional methods have been peculiarly stupid. Learning to read music, to use the voice in singing, and to manipulate the keys of the piano—these put the child through a tiresome drill, eventuating in some measure of ability to sing a song, or to play a piece, but, as a rule, only in a mechanical and uninspired manner. The ability to read and the ability to write music, coupled with the ability to express in tones what one has read or written, are all important features of a musical education, but unless one makes certain that what the child reads and what he plays are a sympathetic rehearsal of what he thinks, his acts are empty of æsthetic content, and difficult and fatiguing in execution. Just as we have improved our methods of verbal reading by making the word rather than the letters the unit of recognition, so in music we should make the melodic phrase rather than the separate tones a basis of instruction. How much more vital and stimulating music-lessons would be, if we could first be shown how to compose simple melodies, how to transcribe them in musical notation, how to play them, and finally how to harmonize them—instead of being constrained to learn scales and five-finger exercises, or to play set pieces to the distracting beats of a metronome! The quite impossible task of holding the attention simultaneously upon the selection of keys, the placing of accents, crescendos, rests, etc., is enough to banish the Muses at the very outset. When one surveys the methods employed in musical education, one wonders no more why so few who "take music-lessons" ever acquire a true appreciation of the art. The wonder is that so many have had the fortitude to persist in their efforts, despite the unsympathetic and subversive methods of their instruction (*109*).

Another serious defect of musical education is the almost universal failure to study musical compositions in the same spirit in which one studies the works of a poet. The best means of gaining familiarity with great music is found in the concert. But here opportunities are limited, and the selections of the program are at best fragmentary. Even if one is able to attend many concerts, one seldom has an opportunity to familiarize oneself with all the important works of any great composer. Many of our teachers of music lack this familiarity, for they have usually been taught by the method of "playing pieces," rather than by any thorough course of study in the classical compositions. In order to attain a scholarly grasp of the field of poetry, it is desirable, at least, to make oneself thoroughly familiar with all the writings of one or two of the greatest poets—such as Dante, Shakespeare, or Goethe. Only in this way do we learn to grasp the inner unity of the spirit which emanates in the appeal of the artist. Is it less important that the scholar in music should know Beethoven, Mozart, or Bach thoroughly? Music has been too largely confined to the individual technique of concert-artists, and we have made the serious mistake of training our pupils as though each of them was destined for a career on the concert-stage. Why should not one be able to read a musical score imaginatively and without the aid of any instrument, just as one reads a poem or a drama? Why should not a musical education primarily aim to develop this imaginative power in the largest number of persons, and allow the more technical features and facilities of communicative expression to occupy the relatively modest place accorded them in the ordinary use of voice and gesture for purposes of normal communication in speech? As a recent writer has aptly expressed it (25), music must be rescued from the "elocutionary" plane upon which it now rests, and placed among the arts which are to be appreciated and understood in the same way as poetry is nowadays so widely appreciated and understood.

Of course it does not follow that ability of expression can be neglected; music, like every other fine art, proceeds from a creative impulse; and unless we learn to exercise our own creative ability, we never can arrive at a true appreciation of the creative work of others. In fact, to listen appreciatively to the music of another requires a high degree of creative ability. The æsthetic attitude of receptivity is never a passive state, and unless one is able to follow a composition with creative expectancy, so as actually to participate in the musical problem and its resolution, one can not be said to enjoy music. All communication involves active response on the part of the recipient, and the hearer of music must both feel and think the music to which he listens. Musical ability never can be developed unless we have first learned to exercise our own musical thought, any more than we can understand the words of another without thinking and feeling in sympathetic response.

While the realization of this end is remote from our present means and aims in education, the need of remedying our scheme in order to give the æsthetic impulse its proper place is now felt in many quarters. One of the most active agents for an improved educational technique is Dalcroze, who, with his method of *eurhythmics,* has already achieved something towards a restoration of the lost art of Greek education. "Educationalists should bear in mind," he tells us (*43,* 148), "that while rhythm plays a preponderant rôle in art, serving to unite all manifestations of beauty, and animating them with the same throbbing life, it should constitute a no less important factor in general education, coordinating all the spiritual and corporal movements of the individual, and evolving in the latter a mental state in which the combined vibrations of desires and powers are associated in perfect harmony and balance." As teachers we must realize that all learning starts with the apprehension of *significant forms* (*10*), whether these be verbal, musical, visual, or manipulatory. Only from these can a clear understanding and a skilful adjustment arise. Yet despite this truth we often try to teach both knowledge and skill by a frontal attack, as though we thought the formal aspects of both must, perforce, accrue quite naturally as a result of our teaching; thus failing to realize that significant form is a condition, as well as a result, of understanding and skilful action.

We resort to a perverse pedagogy whenever we assume that learning is identical with enforced repetition, or that skill is ever achieved through the sheer habituation of practice. There is no learning worthy of the name without insight on the part of the learner, and insight is nothing else than the apprehension of phenomena in their formal or significant completeness. Varying degrees of significance attach to varying degrees of completeness, balance, proportion, and harmony of the formal whole. Education does not start with perfection, but it does start with something that can be perfected by a natural development. As individual agents, we have no control over the conditions of perception. We see and hear what we must see and hear. But to be made to respond with a set exercise which answers no impulsive need of the moment, or, if it does answer, answers imperfectly and without an attendant feeling of satisfaction, is to be forced into a dull routine from which we never can hope to rise masters of the situation. In the education of the ear, and, indeed, in the education of every other avenue of sense, what we need is a more sympathetic approach to the spontaneous springs of action. And these springs of action can be understood, and appropriately aroused and directed, only when the teacher comprehends the significant forms of his subject, knows them as spontaneous motivations, and can so shape the conditions of learning that they will grow through apprehension in completeness and beauty. To do anything well is to do it beautifully; to understand anything well is to comprehend its significant form.

When we rationalize we only abstract the details of a larger pattern and classify them in accordance with certain logical rules of procedure. But there are many ways of classifying, and often it is verbal or numerical symbols with which we are dealing when we suppose ourselves to be examining a concrete experience. Much that passes for education in music is but a facility in the use of tones, either in the analysis of compositions or in the technique of instrumental performance. The spirit of music is entirely lacking, because the significant form of music itself enters neither into the notes of the staff nor into the keys of the pianoforte. So in linguistics, a mechanical study of words is no guarantee that the student will improve his appreciation of poetry or his ability to compose a beautiful line. And hence pedantry is justly criticized for its want of life and the inflexibility of its rules and regulations.

In turning the attention of educators to *rhythm* rather than to *time,* one is emphasizing the importance of life and art, over against the mechanical order which is always impersonal and objective. As some one has observed, "time passes, and is scientifically recorded by the mechanical oscillations of the pendulum. And yet for some of us time 'ambles withal,' for others he 'trots and gallops withal,' for others, again, 'he stands still withal.'"

The significant forms of experience embrace not only

rhythmical movements, but pitch and cadence, intensities, and, last but not least, the "pure design" notable in the effects of visual contours upon contrasting surfaces. These are a few of the outstanding elements which the teacher must come to understand, and upon which he must build his method if he desires to see his pupils enter into the Kingdom of Heaven.

REFERENCES

1. Abraham, O. Töne und Vokale der Mundhöhle. *Zeitschrift für Psychologie*, 1916, *74*, 220-231.
2. Abraham, O. Zur Akustik des Knalles. *Annalen der Physik*, 1919, *60*, 55-76.
3. Abraham, O. Zur physiologischer Akustik von Wellenlänge und Schwingungszahl. *Zeitschrift für Sinnesphysiologie*, 1920, *50*, 121-152.
4. Auerbach, F. *Die Grundlagen der Musik*. Leipzig, 1911, pp. vi + 209.
5. Baird, J. W. Memory for Absolute Pitch. *Studies in Psychology Contributed by Colleagues and Former Students of Edward Bradford Titchener*. Worcester, 1917, pp. 43-78.
6. Baley, S. Versuche über den dichotische Zusammenklang wenig verschiedener Töne. *Zeitschrift für Psychologie*, 1915, *70*, 321-346.
7. Baley, S. Versuche über die Lokalization beim ˙dichotischen Hören. *Zeitschrift für Psychologie*, 1915, *70*, 347-372.
8. Barry, Phillips. Greek Music. *The Musical Quarterly*, 1919, *5*, 578-613.
9. Barton, E. H. *A Text-Book on Sound*. London, 1908, pp. xvi + 687.
10. Bell, Clive. *Art*. New York, [1913], pp. xv + 293.
11. Bernfeld, S. Zur Psychologie der Unmusikalischen. Nebst Bemerkungen über Psychologie und Psychanalyse. *Archiv für die Gesamte Psychologie*, 1915, *34*, 235-253.
12. Beyer, H. Zur Schalleitungsfrage. *Beiträge zur Anatomie usw. des Ohres usw.*, (Passow und Schäfer) 1912, 6, 92-110.
13. Bezold, Fr. and Siebenmann, Fr. *Text-Book of Otology*. Chicago, 1908, pp. xvi + 314.
14. Bing, A. Kritisches zu den Stimmgabelversuchen und deren diagnostischer Wertung. *Archiv für Ohrenheilkunde*, 1915, *96*, 159-182.
15. Bingham, W. V. Studies in Melody. *Psychological Review Monograph Series*, No. 50, 1910, pp. vi + 88.
16. Boring, E. G. and Titchener, E. B. Sir Thomas Wrightson's Theory of Hearing. *American Journal of Psychology*, 1920, *31*, 101-113.

334 REFERENCES

17. Bragg, Sir William. *The World of Sound.* London, 1921, pp. 196.
18. Buck, P. C. *Acoustics for Musicians.* Oxford, 1918, pp. 152.
19. Busoni, F. *Sketch of a New Esthetic of Music.* New York, 1911, pp. 45.
20. Combarieu, Jules. *La Musique, Ses Lois, son Évolution.* Paris, 1911, pp. 348. (Also English translation, New York, 1910.)
21. Curtis, J. Greek Music. *Journal of Hellenic Studies,* 1913, *33,* 33-47.
22. Curtis, Josephine Nash. Duration and the Temporal Judgment. *American Journal of Psychology,* 1916, *27,* 1-46.
23. Dupré, E. et Nathan, M. *Le Langage Musical.* Paris, 1911, pp. vii + 195.
24. Ewald, J. R. Zur Physiologie des Labyrinths. VI. Mittheilung: Eine Neue Hörtheorie. *Archiv für die Gesammte Physiologie* (Pflüger), 1899, *76,* 147-188.
25. Gale, H. Musical Education. *Pedagogical Seminary,* 1917, *24,* 503-514.
26. Gildemeister, M. Untersuchungen über die obere Hörgrenze. *Zeitschrift für Sinnesphysiologie,* 1918, *50,* 161-191.
27. Gray, A. A. *The Labyrinth of Animals.* 2 vols., London, 1907-1908.
28. Halverson, H. M. Binaural Localization of Tones as Dependent upon Differences of Phase and Intensity. *American Journal of Psychology,* 1922, *33,* 178-212.
29. Hanslick, Eduard. *Vom Musikalisch Schönen.* 10te Auf. Leipzig, 1902, pp. viii + 221.
30. Hardesty, I. On the Proportions, Development and Attachment of the Tectorial Membrane. *American Journal of Anatomy,* 1915, *18,* 1-74.
31. Hartridge, H. A Criticism of Wrightson's Hypothesis of Audition. *British Journal of Psychology,* 1921, *12,* 248-252.
32. Hartridge, H. A Vindication of the Resonance Hypothesis of Audition. *British Journal of Psychology,* 1921, 1922. *11,* 277-288; *12,* 142-146; 362-382.
33. Head, Henry and others. Studies in Neurology. 2 vols., London, 1920, pp. xx + 862.
34. Held, H. Zur Kenntnis der peripherer Gehörleitung. *Archiv für Anatomie und Entwickelungsgeschichte,* Jahrgang 1897, 350-360.
35. Helmholtz, H. L. F. *On the Sensations of Tone.* tr. by A. J. Ellis. 3rd edition. London, 1895, pp. xix + 576.
36. Hermann, L. Neue Beiträge zur Lehre von den Vokalen und ihrer Entstehung. *Archiv für die gesammte Physiologie* (Pflüger), 1911, *141,* 1-62.

37. Hermann, L. Phonophotographische Untersuchungen, *Archiv für die gesammte Physiologie* (Pflüger), 1890 and after.

38. Hohenemser, R. Über Konkordanz und Diskordanz. *Zeitschrift für Psychologie*, 1915, *72*, 373-382.

39. Hornbostel, E. M. v. Review: R. M. Ogden, A Contribution to the Theory of Tonal Consonance. *Zeitschrift für Psychologie*, 1912, *61*, 70-71.

40. Hornbostel, E. M. v. Über Vergleichende akustische und musikpsychologische Untersuchungen. *Zeitschrift für angewandte Psychologie*, 1910, *3*, 465-487.

41. Hornbostel, E. M. v. and Wertheimer, M. Über die Wahrnehmung der Schallrichtung. *Sitzungsberichte der preussischen Akademie der Wissenschaften*, 1920, 388-396.

42. Jaensch, E. R. Die Natur der menschlichen Sprachlaute. *Zeitschrift für Sinnesphysiologie*, 1913, *47*, 219-290.

43. Jaques-Dalcroze, Émile. *Rhythm, Music and Education*. London, 1921, pp. xv + 257.

44. Jespersen, O. *Phonetische Grundfragen*. Leipzig und Berlin, 1904, pp. iii + 185.

45. Kemp, W. Methodisches und Experimentelles zur Lehre von den Tonverschmelzung. *Archiv für die gesamte Psychologie*, 1913, *29*, 139-257.

46. Klemm, O. Untersuchungen über die Lokalization von Schallreizen. 3te Mitteilung. *Archiv für die gesamte Psychologie*, 1918, *38*, 71-114.

47. Klemm, O. Untersuchungen über die Lokalization von Schallreizen. 4te Mitteilung. *Archiv für die gesamte Psychologie*, 1920. *40*, 117-146.

48. Köhler, W. Akustische Untersuchungen I, II. *Zeitschrift für Psychologie*, 1910, *54*, 241-289; *58*, 59-140.

49. Köhler, W. Akustische Untersuchungen III und IV (vorläufige Mitteilung). *Zeitschrift für Psychologie*, 1913, *64*, 92-105.

50. Köhler, W. Akustische Untersuchungen, III. *Zeitschrift für Psychologie*, 1915, *72*, 1-192.

51. Krehbiel, H. E. *Afro-American Folk Songs*. New York, 1914, pp. xii + 176.

52. Krüger, F. Die Theorie der Konsonanz. *Psychologische Studien*, 1910, *5*, 294-411.

53. Kühn, W. Experimentelle Untersuchungen über das Tonalitätsgefühl. *Beiträge zur Anatomie usw. des Ohres usw.*, 1919, *13*, 254-278.

54. Lachmund, H. Vokal und Ton. *Zeitschrift für Psychologie*, 1921, *88*, 1-52.

55. Lanier, Sidney. *The Science of English Verse*. New York, 1888, pp. 315.

56. Lantier, —. Le traitement des surdités de guerre. *Comptes rendus academie des sciences*, 1917, *164*, 419-421.

57. Liebermann, P. v. und Révész, G. Die binaurale Tonmischung. *Zeitschrift für Psychologie*, 1914, *69*, 234-255.

58. Lipps, Theodor. *Psychologische Studien*. Leipzig, 1905, pp. 115 ff.

59. Maltzew, Catharina v. Das Erkennen sukzessiv gegebener musikalischer Intervalle in den äusseren Tonregionen. *Zeitschrift für Psychologie*, 1912, *64*, 161-257.

60. Marage, —. Contribution a l'étude de consonnes. *Comptes rendus academie des sciences*, 1911, *152*, 1265-1267.

61. Marage, —. Contribution a l'étude des hypoacousies consécutives à des blessures de guerre. *Comptes rendus academie des sciences*, 1915, *161*, 148-150.

62. Marage, —. Synthèse des voyelles. *Comptes rendus academie des sciences*, 1900, *130*, 746-748.

63. Marage, —. Traitement des hypoacousies consécutives à des blessures de guerre. *Comptes rendus academie des sciences*, 1915, *161*, 319-322.

64. Meyer, Max F. An Introduction to the Mechanics of the Inner Ear. *University of Missouri Studies*, 1907, vol. II, no. 1, pp. vi + 140.

65. Meyer, Max F. Contributions to a Psychological Theory of Music. *University of Missouri Studies*, 1901, vol. I, no. 1, pp. vi + 80.

66. Meyer, Max F. On the Attributes of Sensation. *Psychological Review*, 1904, *11*, 83-103.

67. Meyer, Max. F. Über Tonverschmelzung und die Theorie der Consonanz, *Zeitschrift für Psychologie*, 1898, *17*, 404-421.

68. Meyer, Max F. Vorschläge zur akustischen Terminologie. *Zeitschrift für Psychologie*, 1914, *68*, 115-123.

69. Meyer, Max F. Zur Theorie japanischer Musik. *Zeitschrift für Psychologie*, 1903, *33*, 289-306.

70. Miller, Dayton C. *The Science of Musical Sounds*. New York, 1916, pp. viii + 286.

71. M'Kendrick, J. G., and Gray, A. A. The Ear. *Text-Book of Physiology*. E. A. Schäfer, ed. 1900, vol. II, pp. 1149-1205.

72. Monro, D. B. *The Modes of Ancient Greek Music*. Oxford, 1894, pp. xvi + 144.

73. Moore, H. T. The Genetic Aspect of Consonance and Dissonance. *Psychological Review Monograph Series*, No. 73, 1911, pp. 68.

74. Myers, C. S. Abstract: Theories of Consonance and Dissonance. Proceedings of the Psychological Society. *British Journal of Psychology*, 1905, *1*, 315-316.

75. Myers, C. S.   A Case of Synæsthesia.   *British Journal of Psychology*, 1911, *4*, 228-238.

76. Myers, C. S.   The Beginnings of Music.   *Essays and Studies Presented to William Ridgeway.*   Cambridge, 1913, 579-582.

77. Myers, C. S.   Two Cases of Synæsthesia.   *British Journal of Psychology*, 1914, *7*, 112-117.

78. Ogden, R. M.   A Contribution to the Theory of Tonal Consonance. *Psychological Bulletin*, 1909, *6*, 297-303.

79. Ogden, R. M.   Eurhythmic.   *Sewanee Review*, 1920, *28*, 520-543.

80. Ogden, R. M.   Summaries on Hearing.   *Psychological Bulletin*, 1911-1920, *8-17*, 93-100; 116-123; 107-116; 96-105; 161-169; 189-197; 159-164; 76-85; 142-148; 228-238.

81. Ogden, R. M.   The Attributes of Sound.   *Psychological Review*, 1918, *25*, 227-241.

82. Ogden, R. M.   The Tonal Manifold.   *Psychological Review*, 1920, *27*, 136-146.

83. Parker, G. H.   A Critical Survey of the Sense of Hearing in Fishes.   *Proceedings of the American Philosophical Society*, 1918, *57*, 69-98.

84. Parry, C. H. H.   *The Evolution of the Art of Music.*   New York, 1912, pp. x + 342.

85. Patterson, W. M.   *The Rhythm of Prose.*   New York, 1916, pp. xxiii + 193.

86. Pear, T. H.   The Classification of Observers as "Musical" and "Unmusical."   *British Journal of Psychology*, 1911, *4*, 89-94.

87. Pear, T. H.   The Experimental Examination of Some Differences between the Major and the Minor Chords.   *British Journal of Psychology*, 1911, *4*, 56-88.

88. Plato.   *Philebus*, 56.

89. Pohlman, A. G.   Abstracts: The Problem of Middle-Ear Mechanics and its Relation to Sound Transmission and Sound-analysis.   *The Anatomical Record*, 1920, *18*, 253-4.   A Note on the Relation of the Auricle and External Auditory Drum-Membrane Mechanics.   *Ibid.*, 1921, *21*, 76.

90. Pratella, Balilla.   *Musical Courier*, 1915, *71*, 5 ff.

91. Pratt, C. C.   Bisection of Tonal Intervals Smaller than an Octave. *Journal of Experimental Psychology*, 1923, *6*, 211-222.

92. Pratt, C. C.   Some Qualitative Aspects of Bitonal Complexes. *American Journal of Psychology*, 1921, *32*, 490-515.

93. Raman, C. V.   On the Maintenance of Combinational Vibrations by two Simple Harmonic Forces.   *Physical Review*, 1915, *5*, 1-20.

94. Ranjard, —.   Contribution à l'étude de l'audition et de son développement par les vibrations de la sirène à voyelles.   *Comptes rendus academie des sciences*, 1910, *150*, 724-726.

95. Ranjard, —. Sur les cent premiers cas des surdité traités par la méthode de Marage au Centre de rééducation auditive de la 8ᵉ région. *Comptes rendus academie des sciences,* 1917, *163,* 243-245.

96. Révész, G. *Zur Grundlegung der Tonpsychologie.* Leipzig, 1913, pp. viii + 148.

97. Rich, G. J. A Study of Tonal Attributes. *American Journal of Psychology,* 1919, *30,* 121-164.

98. Rupp, H. Über die Prüfung musikalischer Fähigkeiten. *Zeitschrift für angewandte Psychologie,* 1914, *9,* 1-76.

99. Schäfer, K. L. Der Gehörssinn. *Handbuch der Physiologie des Menschen.* Herausgegeben von W. Nagel. Braunschweig, 1905, Bd. III. pp. 476-588.

100. Schäfer, K. L. Über die Wahrnehmbarkeit von Kombinationstönen bei partiellem oder totalem Defekt des Trommelfelles. *Beiträge zur Anatomie usw. des Ohres usw.* (Passow und Schäfer), 1913, *6,* 207-218.

101. Schäfer, K. L. Untersuchungsmethodik der akustischen Funktionen des Ohres. *Handbuch der Physiologischen Methodik,* (Tigerstedt) Leipzig, 1914, vol. III, Part IIIb, 204-394.

102. Schole, H. Über die Zusammensetzung der Vokale, U, O, A. *Archiv für gesamte Psychologie,* 1918, *38,* 12-70.

103. Schulze, F. A. Die Schwingungsweise der Gehörknöchelchen. *Beiträge zur Anatomie usw. des Ohres usw.* (Passow und Schäfer), 1911, *4,* 161-165.

104. Seashore, C. E. Avocational Guidance in Music. *Journal of Applied Psychology,* 1917, *1,* 342-348.

105. Seashore, C. E. The Measurement of Musical Talent. *The Musical Quarterly,* 1915, *1,* 129-148.

106. Seashore, C. E. *The Psychology of Musical Talent.* Boston, 1919, xvi + 288.

107. Seashore, C. E. The Tonoscope. *Psychological Review Monograph Series,* No. 69, 1914, 157-160.

108. Seashore, C. E. and Mount, G. H. Correlation of Factors in Musical Talent and Training. *Psychological Review Monograph Series,* No. 108, 1918, 47-92.

109. Seymour, Harriet A. *How to Think Music.* New York, 1910, pp. 52.

110. Shambaugh, G. E. Die Frage der Tonempfindung. *Archiv für die gesammte Physiologie* (Pflüger), 1911, *138,* 155-158.

111. Spalding, W. R. *Music: An Art and a Language.* Boston, 1920, pp. 342.

112. Sterzinger, O. Rhythmische Ausgeprägtheit und Gefälligkeit musikalischer Sukzessivintervalle. *Archiv für die gesamte Psychologie,* 1916, *35,* 75-124.

113. Sterzinger, O. Rhythmische und ästhetische Charakteristik der musikalischen Sukzessivintervalle und ihre ursächlichen Zusammenhänge. *Archiv für die gesamte Psychologie,* 1917, *36,* 1-58.

114. Stewart, G. W. Binaural Beats. *Psychological Review Monograph Series,* No. 108, 1918, 31-46.

115. Stewart, G. W. The Function of Intensity and Phase in the Binaural Location of Pure Tones. *Physical Review,* 1920, *15,* 425-445.

116. Stumpf, C. Beobachtungen über Kombinationstöne. *Zeitschrift für Psychologie,* 1910, *55,* 1-142.

117. Stumpf, C. *Die Anfänge der Musik.* Leipzig, 1911, pp. 209.

118. Stumpf, C. Die Struktur der Vokale. *Sitzungsberichte der preussischen Akademie der Wissenschaften,* 1918, Erster Halbband, 333-358.

119. Stumpf, C. Differenztöne und Konsonanz (Zweiter Artikel). *Zeitschrift für Psychologie,* 1911, *59,* 161-175.

120. Stumpf, C. Geschichte des Consonanzbegriffes. Erster Teil. *Abhandlungen der philosophisch-philologische Classe der Königl.-Bayerischen Akademie der Wissenschaften,* 1901, *21,* 1-78.

121. Stumpf, C. Konsonanz und Konkordanz. *Zeitschrift für Psychologie,* 1911, *58,* 321-355.

122. Stumpf, C. *Tonpsychologie.* Leipzig: I, 1883, pp. xivl+427; II, 1890, pp. xiv + 582.

123. Stumpf, C. Über neuere Untersuchungen zur Tonlehre. *Bericht über den VI. Kongress für experimentelle Psychologie in Göttingen.* Leipzig, 1914, 305-348.

124. Stumpf, C. Veränderungen des Sprachverständnisses bei abwärts fortschreitender Vernichtung der Gehörsempfindung. *Beiträge zur Anatomie usw. des Ohres usw.* (Passow und Schäfer), 1921, *17,* 182-190.

125. Stumpf, C. Zur Analyse der geflüsterter Vokale. *Beiträge zur Anatomie usw. des Ohres usw.* (Passow und Schäfer), 1919, *12,* 234-254.

126. Stumpf, C. Zur Analyse der Konsonanten. *Beiträge zur Anatomie usw. des Ohres usw.* (Passow und Schäfer), 1921, *17,* 151-181.

127. Stumpf, C. u. Hornbostel, E. M. v. Über die Bedeutung ethnologischer Untersuchungen für die Psychologie und Aesthetik der Tonkunst. *Bericht über den IV. Kongress für experimentelle Psychologie in Innsbruck.* Leipzig, 1911, 256-269.

128. Surette, Thomas Whitney. *Music and Life.* Boston, 1917, pp. xvi + 251.

129. Valentine, C. W. The Method of Comparison in Experiments with Musical Intervals and the Effect of Practice on the Apprecia-

tion of Discords. *British Journal of Psychology*, 1914, *7*, 118-135.

130. Watson, F. R. An Investigation of the Transmission, Reflection, and Absorption of Sound by Different Materials. *Physical Review*, 1916, *7*, 125-132.

131. Watson, J. B. *Psychology from the Standpoint of a Behaviorist.* Philadelphia, 1919, pp. xiv + 429.

132. Watt, H. J. A Theory of Binaural Hearing. *British Journal of Psychology.* 1920, *11*, 163-171.

133. Watt, H. J. *The Foundations of Music.* Cambridge, 1919, pp. xvi + 239.

134. Watt, H. J. *The Psychology of 'Sound.* Cambridge, 1917, pp. viii + 241.

135. Watt, H. J. The Typical Form of the Cochlea and Its Variations. *Proceedings of the Royal Society*, 1916, *89*, 410-421.

136. Weiss, A. P. The Tone Intensity Reaction. *Psychological Review*, 1918, *25*, 50-80.

137. Weld, H. P. An Experimental Study of Musical Enjoyment. *American Journal of Psychology*, 1912, *23*, 245-308.

138. Wien, M. Über die Empfindlichkeit des menschlichen Ohres für Töne verschiedener Höhe. *Archiv für die gesammte Physiologie* (Pflüger), 1903, *97*, 1-57.

139. Wittmaack, —. Eine neue Stütze der Helmholtz'schen Resonanz Theorie. *Archiv für die gesammte Physiologie* (Pflüger), 1907, *120*, 249-252.

140. Wittmaack, K. Vergleichende Untersuchungen über Luftschall—Luftleitung und Bodenschwingung—Körperleitungschädigungen des akustischen Apparates. *Archiv für Ohren-, Nasen-, und Kehlkopfheilkunde*, 1918, *102*, 96-107.

141. Wittmaack, K. Zur Kenntins der Cuticulargebilde des inneren Ohres mit besonderer Berücksichtigung der Lage der Cortischen Membran. *Jenaische Zeitschrift für Naturwissenschaft*, 1918, *55*, 537-573.

142. Wittmann, J. Neuer objektiver Nachweis von Differenztönen erster und höherer Ordnung. *Archiv für die gesamte Psychologie*, 1915, *34*, 277-315.

143. Wolff, H. J. Monochord und Stimmgabeluntersuchung zur Klärung der Beziehung zwischen Luft und Knochenleitung bei Normal- und Schwerhörigen. *Beiträge zur Anatomie usw. des Ohres usw.* (Passow und Schäfer), 1911, *5*, 131-150.

144. Wood, Alexander. *The Physical Basis of Music.* Cambridge, 1913, pp. vi + 163.

145. Wrightson, T. *An Enquiry into the Analytical Mechanism of the Internal Ear.* London, 1918, pp. xi + 254.

146. Wundt, Wilhelm. *Völkerpsychologie: Die Sprache.* 2 vols., 2nd ed., Leipzig, 1904, pp. xv + 667; x + 673.

147. Yoshii, —. Experimentelle Untersuchungen über die Schädigung des Gehörorgans durch Schalleinwirkung. *Zeitschrift für Ohrenheilkunde,* 1909, *58,* 201 ff.

148. Zahm, J. A. *Sound and Music.* Chicago, 1900, pp. 452.

149. Zimmermann, Gustav. *Die Mechanik des Hörens und Ihre Störungen.* Wiesbaden, 1900, pp. viii + 110.

150. Zwaardemaker, H. Über Hörapparate. *Archiv für Ohrenheilkunde,* 1919, *104,* 1-38.

INDEX

343

INDEX

INDEX

INDEX

INDEX

INDEX

INDEX

INDEX

INDEX

INDEX

INDEX

Dullness, 58, 78, 82, 102, 113, 115, 118, 119, 166.
Dupré, 295.
Duration, 61-67, 69, 70, 72, 77, 78, 98, 105, 107, 114, 116, 117, 134, 166, 168, 169, 208-210, 213, 221, 232, 258, 264, 269, 275, 276, 316, 317, 322.

Ear-drum, 23, 26, 27, 29, 39, 277, 285.
Ear-trumpets, 291.
East Indian music, 198, 233.
Echo, 5.
Ellis, 184 n., 196.
Endolymph, 33, 39.
Enharmonic music, 236, 238.
Enharmonic steps, 187.
Enharmonic tetrachord, 179.
Equal division, 187, 195, 198, 200, 201, 234, 235.
Equal intervals, 122, 144, 156-158, 165, 168, 173, 174, 176, 191, 196, 197, 199, 200, 226, 228, 234, 236, 239, 253, 322.
Equal temperament, 157, 173, 193, 195-197.
Equilibration, 25, 104.
Equilibrium, 23-25, 104.
Eurhythmics, 303, 329.
Euripides' *Orestes,* 177.
Eustachian tubes, 27, 284, 287.
Ewald, 43, 44.
Extensity, 62, 66, 67, 77, 311, 316.
External auditory meatus, 25, 26, 274, 276, 282.
External ear, 274, 275.

Five-tone scale, 233.
Folk-music, 190, 240, 243, 308, 320.
Formant, 80-83, 85, 86, 88-102, 107, 109-111, 113-116, 118, 119, 144, 146, 148, 166, 167, 206, 208, 215-217, 221, 280, 288, 289.
Fourier analysis, 84, 94.

Fourier's law, 41.
Fourth from the bass, 187, 249.
Frequency-number, 90, 91.
Frequency of vibration, 4, 9, 12-21, 35-39, 43, 44, 46, 50, 51, 58, 60, 61, 66, 77, 78, 86-90, 93-95, 98, 103, 109, 112, 114, 115, 119, 250, 267, 293.
Frequency-ratio, 250.
Frequency-tone, 90, 91, 94.
Fundamental, 13-16, 39, 80-85, 87-92, 94, 96, 97, 114, 128, 131, 152, 184, 215, 217, 249, 261, 320.
Fusion, 123-130, 133-150, 159, 161, 170, 174, 183, 188, 226, 230-232, 244-246, 248, 250, 291, 320.
"Futurism," 201, 236, 238, 239.

Gale, 328.
Generators, 4, 5, 12, 13, 18-20, 40, 53, 86, 103, 112, 128, 131, 132, 206.
Gildemeister, 278.
Goethe, 327.
Grace-notes, 172.
Gray, 25.
Greek education, 303, 308, 329.
Greek enharmonics, 179.
Greek mode, 182.
Greek music, 176, 177, 180, 191.
Greek scale, 174, 178, 183-187, 189, 231.
Gregorian music, 187, 218.

Halverson, 265, 266, 270.
Hammer, 27.
Hanslick, 255.
Hardesty, 42.
Harmonic conformity, 245, 246.
Harmonic theory, 140, 146, 148.
Harmonics, 80, 88, 90, 98, 141, 143, 146, 148, 149, 152, 153, 156, 157, 159, 163, 164, 166, 173-178, 181, 183, 184, 186, 187, 190, 191, 193, 195, 198, 200, 228, 229, 235, 236, 238, 244.